The Object of My Infection

The Object of My Infection

৵৽

Tara Hulko

THE OBJECT OF MY INFECTION

ISBN: 978-0-557-06581-3

Dedication

I wanted to personally give extra special Love & Thanks to Lauren, Jason, Terra, Shandy, Carla, Sara, Trish, Linda, Heidi, and Renee for their strength and perseverance. Thank you all for sharing your stories with the world.

And to all of us who have been effected with Lyme disease. A community of some of the kindest and strongest people I have ever known.

Together we will all help find a cure

Prologue

One signer of the Declaration of Independence, Dr. Benjamin Rush, who was also George Washington's physician, predicted the following. "Unless we put medical freedom into The Constitution, the time will come when medicine will organize into an undercover dictatorship."

Would you like to see ignorance at its best? Try telling your physician that you have Lyme disease. All the symptoms are there, screaming out loud. You were bitten by a tick, presented with a bulls eye shaped rash; you are suddenly losing your ability to think clearly, having unexplained heart complications, tremors and seizures. Your body severely aches like it never has before and the fatigue is unending.

However, despite having the notorious
" bull' s eye rash" , your Lyme disease blood
test came back negative. Welcome to the
wonderful world of Lyme disease.

Who would have ever thought that a type
of bacteria would be the controversy that it is.
Many of us " Lymies" fight every single day
to get proper care. The fact that doctors argue
over the proper protocol and insurance
companies are denying what they consider
unnecessary treatments is an ongoing battle.
Amazing doctors are losing their licenses while
healing patients; having their methods called
unconventional and extreme. Having to ask
around to find any " Lyme Literate" doctors,
because the physicians are too afraid to
advertise their services in fear of the pressures
that would lie ahead. Having your disease be
treated like early years of the AIDS explosion.
Being denied treatment, because the
government lacks proper diagnostic testing and
Lyme conditions go undiagnosed until a victim
falls so deeply ill that they suffer chronic
affects for the rest of their lives. You would
never know a " Lymies" life, until your friend
or family member is afflicted with this. This is
the silent epidemic that no one likes to speak

about. Have you ever asked yourselves why? I mean, who really benefits from our suffering?

Our doctors have been force fed by the IDSA (Infectious Disease Society of America), that Lyme disease is a mild infection that can be easily cured with a 2 week course of oral antibiotics. Most of us battling advanced Lyme know this is in fact not the case. Sure, plenty of us know of someone who had Lyme disease and got better without a problem, but many of us were not that lucky. This is due to the fact there are two forms of Lyme disease and over 300 strains– approximately 100 of which are in the United States alone. There is even debate whether there is possibly a viral strain of Lyme.

Within this journey, I want to hold your hand and show you our story. I want to show you the truth behind Lyme disease. This is a world corrupted by the CDC, IDSA and pharmaceutical companies. A world where the sick get sicker. We have been told that we suffer from forms of mental illness or we are just looking for attention. We shell out thousands upon thousands of dollars to receive proper treatment because our insurance companies deny our care. We have been told we have an illness that doesn' t exist except in our psychosomatic minds. We have been told that we want to be sick. That we want to be a burden on those that we love and we want to lose the productive lives that we once lived- that we all dream we can have again. I will open your eyes to a new world that will crack open the secrets; so we can all shout out that the Lyme community should not have to be treated like this anymore. We will point the finger at those responsible in the controversy and make them own up and answer to the questions on why they have turned a blind eye on an epidemic that is infecting a possible 200,000 people a year.

Chapter 1

Getting to Know You

So this is the part of the book where I recite to you the life story of Lyme disease and how it came to be. Most of this information, every " Lymie" could recite while standing on their heads. Actually, I stand corrected. We should pretend we could, because if it was actually possible for us to stand on our heads, that would cause a whole world of trouble! However, there are a few lesser known facts about this disease that have been discovered.

Borrelia burgdorferi, or better known as Lyme disease, is a type of Spirochete. Spirochetes are spiral shaped bacterium which are long and thin. Syphilis is another type of spirochete and close relative to Borrelia burgdorferi (*Bb*). Having a cork screw shape allows the bacteria to drill its way into all bodily tissues and organs. This is a defense mechanism to help it hide from medications and diagnostic testing.

The bacterium is thousands of times larger than a virus. However, it still requires a powerful microscope to see one. Roughly 1,500 *Bb* must be laid end to end to equal one inch. About 100,000 of *Bb* standing side to side would equal one inch.

It is a popular misconception that Lyme disease was discovered in the late 1970's in Lyme, Connecticut. However, medical literature in Europe, is documented with more than a century of acknowledgement about the condition.

The first record of a condition associated with Lyme disease dates back to 1883 in Breslau, Germany, where a physician named Alfred Buchwald described a degenerative skin disorder now known as Acrodermatitis chronica atrophicans (ACA). Then in 1909, a physician named Arvid Afzelius presented research about an expanding, ring like lesion he had observed. 12 years later, Afzelius speculated in his notations, that the rash came from the bite of an *Ixodes* tick (a type of hard bodied ticks).

Throughout the early 1900' s, associations were being made among many of the symptoms and signs that constitute Lyme disease. Some of these associations were: joint involvement in patients with late disease (1921), the link between the EM (Erythema Migrans or bulls eye) rash and neurologic problems (1922), psychiatric symptoms in patients with the EM rash (1930), patients with benign lymphocytomas observed to also have either EM or ACA (1934), and the description of heart involvement that appeared in patients with both the EM rash and arthritic symptoms (1934). By mid-century, physicians were experimenting with still-novel antibiotics and reporting successful results.

In 1970, for the first time, an incidence of EM known with certainty to have been acquired in the United States was reported by Rudolph Scrimenti, who diagnosed and treated a patient who had been bitten by a tick while hunting grouse in Wisconsin and acquired the disease. Later in 1975, he publishes his treatments of the rash with Penicillin.

In 1976, the first US case of clustering of this disease was reported in a Navel Medical Hospital in Southwestern Connecticut.

In 1977, physician Allen Steere described the first clustering of the disease misdiagnosed as juvenile rheumatoid arthritis. They named this " new" condition 'Lyme arthritis' after the explosion of it coming from Lyme, Connecticut.

In the early 1980's, an entomologist at the United States Rocky Mountain Laboratories of the National Institutes of Health by the name of Willy Burgdorfer, MD, Ph.D., was investigating outbreaks of Rocky Mountain spotted fever. Burgdorfer noticed an embryonic form of parasite in the body fluid of two of the ticks. Guided by his extensive knowledge of the early scientific writings of European researchers, he undertook a very close inspection of the tick and found poorly stained, sluggish spirochetes. Within a year, the spirochetes had been named *Borrelia burgdorferi (Bb)*, in his honor, and definitely identified as the causative agent of Lyme disease. Dr. Burgdorfer was the partner in the successful effort to culture the spirochete, along with Alan Barbour, MD.

In 1985, Paul Duray, a Lyme disease researcher, declared that the Lyme disease bacterium scatters itself throughout the body early in the course of infection. Also in 1985, Burgdorfer was able to demonstrate that ticks infected with the Lyme spirochete could be found across the country. [1]

The most common carrier of the disease is the deer tick, however all ticks are carriers. The tick can be black or brown and as small as a pin head in its Nymph stage. Due to this reason, most infected individuals never realized they were bitten. Ticks are everywhere.

Transmission of Borrelia Burgdorferi enters the body immediately and has been found in the central nervous system as early as 24 hours after the initial bite. It prefers to move directly into the tissue where it can move easily and attack the organs and bodily systems. *Bb* will grow in cycles of 4-6 weeks. If the bacteria are not fully killed off within the body in the beginning, it will grow again during its next cycle. This is why initial treatment recommended by the CDC of only 2 weeks is usually ineffective if the Lyme has not been diagnosed the day of the infection. This usually only works for people who noticed a bite and went to the doctor the right away.

In later stages, Bb, the causative agent of Lyme disease is pleomorphic. This means that it can mutate into several different forms and as a result, is able to evade the immune system, antibiotics and can make diagnostic testing very difficult. It can go into hiding and lie dormant for varied periods of time.

Currently it is estimated that there are over 200,000 new cases of Lyme disease every year. Unfortunately it is also estimated that only 20,000 cases are reported to the CDC. It is required to have a positive ELISA and Western Blot test to fit into the CDC's criteria. While the standard issued ELISA test currently tests for only 1 strain, many strains go undetected for this reason. Currently there is only one lab that tests for 2 different strains. Most Lyme Literate doctors will only test through this lab since they specialize in tick borne infections, however most insurance companies will not cover the cost of testing done at this lab.

Having a negative titer does not mean that you are negative for Lyme. There are many reasons for a false test. For example, it is a common known fact that the medical community recognizes the problem of pleomorphism and the resultant difficulty with diagnostic testing when it occurs. However, when it comes to Lyme disease, astonishingly the medical community feels that this rule does not apply.

Another known phenomenon that occurs allowing testing to fail is if a patient is co-infected with a very common tick borne disease called Bartonella. Bartonella, also known as cat scratch fever, has the unique ability to turn off the bodies ability to produce antibodies, so when doctors rely on tests that only produce positive results when antibodies are present, they are not only unable to diagnose Bartonella, they can no longer diagnose Lyme or any other of the many types of Tick co-infections that come packaged with Lyme disease.

These are just some examples of why most people go undiagnosed when they fail the traditional ELISA blood screening. The ELISA test has been proven to be unreliable with only a 50% accuracy rating. These tests were designed by the CDC and we NOT meant to be a diagnostic test and the fail rate is very high. The CDC has this posted on their website.

Lyme has always meant to be a clinical diagnosis and therefore should only diagnosed by specialists in this field.

Chapter 2

What the Hell is Wrong with Me???

Right now or possibly one day, you may question if you have Lyme disease. Maybe you were in the woods hiking or camping, perhaps you were gardening. What if you were vacationing in a highly infested area? Could it be that you have pets or you saw circular rash on your body. Or maybe, just maybe, you are extremely sick and no one can seem to explain why.

You might be asking yourself- what should I be looking for or what is it like? Lyme is called " The New Great Imitator" . It can pretty much attack or mimic many illnesses making a clinical diagnosis even difficult. Ahead I will lay out the <u>most common</u> information.

There are two different forms of Lyme disease. Early Localized and Disseminated Lyme. Depending on what type or stage you are in, depends a lot on how your road of treatment will progress.

<u>Lyme disease Symptoms</u>

<u>EARLY LOCALIZED DISEASE (Early stage or weak strain- easy to treat)</u>

Symptoms: Headache, stiff neck, fever, muscle aches, and fatigue. Usually felt as flu-like symptoms.*

Less then 50% of fair skinned patients a unique enlarging rash, referred to as Erythema Migrans (EM), days to weeks after the bite. On dark-skinned people, this rash resembles a bruise. This rash may start as a small, reddish bump about one-half inch in diameter. It may be slightly raised or flat. It soon expands outward, often leaving a clearing (normal flesh color) in the center. It can enlarge to the size of a thumb print or cover a person's back.

To be considered local disease, the rash must be at the tick bite site with no other major organ system involvement. A rash occurring at a site other then the bite site is an indication of Disseminated Lyme Disease.

DISSEMINATED LYME DISEASE (Advanced Lyme or Strong Strain)

Some people do not notice the early indications of infection. Early manifestations usually disappear, and disseminated (other organ system involvement) infection may occur. General symptoms alone do not indicate Disseminated Lyme Disease.

Symptoms: Profound fatigue, severe headache, nausea, fever(s), severe muscle aches/pain, stiff neck, meningitis, vision changes, seizures, Bells Palsy, cognitive changes, sleep disorders, vertigo, bulls eye rash, heart complications, severe pain in the joints, kidney and liver abnormalities, swelling of the lymph nodes*

Other symptoms may and can occur, refer to a Lyme Literate Doctor for a true diagnosis.

There are a few medical conditions peak the interest of a Lyme Doctor (LLMD). If you have ever been told you have the following, you may want to just follow it up with a Lyme check. Fibromyalgia (FM), Chronic Fatigue Syndrome (CFS), Gulf War Illness, interstitial cystitis, Irritable Bowel Syndrome (IBS), thyroid complications, non-epileptic seizures, Chronic Epstein Barr Virus (CEBV), Bi-polar disorder, psychosis, schizophrenia, multiple sclerosis (MS), rheumatoid arthritis, lupus, or other autoimmune and neurodegenerative diseases.

30-50% of acute Lyme disease patients went on to develop chronic Lyme disease. Additionally, some previously asymptomatic patients may reactivate their infection following various stressors such as trauma, surgery, pregnancy, coexisting illness, antibiotics treatment, or severe psychological stress. The Lyme vaccine also did reactivate infections. This is not a new phenomenon with infectious diseases. [2]

Depending on which side of the fence your doctor is on, will strongly dictate your treatment. Most general practitioners will opt to go by the general guidelines of the CDC and offer you an antibiotic such as Doxycycline for 14-21 days. This is fine if you have localized Lyme or a new infection, otherwise you will be in a world of trouble soon.

If you have a disseminated strain or a previously undiagnosed case, then treatment with long term antibiotic therapy is a must. If given the option of IV therapy, you are looking at a minimum of 3 months of medication. If your insurance will not cover the treatment and this does happen quite often, then you are left to battle this on oral antibiotics. It is very important that you are given a long course of treatment due to the growth cycles of the Lyme disease otherwise it will replicate and all the prior progress will be in vain. In addition, if it is not properly killed of it will create a super bug or go into a protective mode allowing itself to become unaffected by antibiotics and you will end up with chronic Lyme disease.

If you or someone you love is ever questioning the possibility of Lyme, I urge you to immediately go to a Lyme Literate doctor from the first step. With any major disease, you are always better off seeking a specialist. An infectious disease doctor does not immediately classify the doctor as a Lyme specialist. There are many people and resources online willing to recommend the best in every state. When time is of the essence in whether a small course of medication is an option, best to do it right the first time!

Chapter 3

Shame on You! What Have You Done????

The Infections Disease Society of America (IDSA) represents physicians, scientists and other health care professionals who specialize in infectious diseases. Their purpose is to write treatment guidelines, to improve the healthcare system by promoting education, research and prevention of infectious diseases.

In 2000 and 2006, the IDSA released the guidelines on Lyme disease. Gary P. Wormser, Raymond J. Dattwyler, Eugene D. Shapiro, John J. Halperin, Allen C. Steere, Mark S. Klempner,

Peter J. Krause, Johan S. Bakken, Franc Strle, Gerold Stanek, Linda Bokenstedt, Durland Fish, J. Stephen Dumler, and Robert B. Nadelman are the following names belonging to the individuals who made up these guidelines. These are the names of the individuals who have caused all the controversy over a type of bacteria.

In these guidelines, the IDSA has taken the stance that Lyme disease is rare, hard to get, easy to treat and there is no such cases of chronic infections, other transmissions such as sexually or via mother to placenta.

Upon reviewing the IDSA' s, Dr. Eugene Shapiro' s online profile for Yale Medical at Yale University, he is quoted with saying," The anxiety is as large a problem, or larger, than Lyme disease itself," he says. " There is a lot of misinformation in the lay press and on the Internet, and misdiagnosis is rampant." He continues to go on that, " Lyme disease is fairly easy to diagnose and cure. But the improper use of diagnostic tests, which can give false positive results, has led to over diagnosis. Some symptoms, such as fatigue, headaches and chronic pain, are widespread and most often are due to any number of other causes, so (Shapiro) tests for it only when objective findings suggest that someone has it. Those findings include an inflamed knee joint or facial palsy. " Adding, " I have an additional problem as a pediatrician because this is the perfect disease for parental paranoia," he says. Although some parents are under the impression that Lyme disease can ruin their child' s health, in reality he says that about 90 percent of all cases consist of a simple rash."

I think his words speak for themselves about how off the mark this individual is and how dangerous it was to allow him to be one of the IDSA' s panel members deciding for all doctors the proper way to diagnose and treat Lyme disease.

According to the IDSA' s guidelines, if a patient goes in with a tick bite, a single dose of Doxycycline may be administered at 200mg for an adult if residing in New England, parts of the Mid-Atlantic States, parts of Minnesota and Wisconsin, but not in other locations of the United States. Rates of Bb infection are low in almost the entire region in which the tick is epidemic. Testing of ticks for tick borne infections is not recommended, except for the purpose of research studies.

Patients are to be monitored for 30 days after a tick bite to watch for bull' s eye rash at the site of the bite. ONLY If patient presents with a EM rash and a positive ELISA (Enzyme-linked immunosorbent assay) test and Positive Western Blot test, they are to be administered 100mg of Doxycycline twice a day or amoxicillin (500 mg 3 times per day), or cefuroxime axetil (500 mg twice per day) for 14 days (10– 21 days for Doxycycline and 14– 21 days for amoxicillin or cefuroxime axetil) is recommended for the treatment of adult patients with early localized or early disseminated Lyme, in the absence of specific neurological manifestations. With Lyme meningitis and other manifestations of early neurological Lyme disease, use ceftriaxone (2g once per day intravenously for 14 days) in early Lyme disease is recommended for adult patients with acute neurologic disease manifested by meningitis or radiculopathy. Late stage Lyme Disease may be treated up to 8 weeks of oral antibiotics or 2–4 weeks if IV antibiotics.

Now what I want to know is what other major late stage infectious disease has ever been cured in 2–4 weeks of oral antibiotics? AIDS?? Syphilis??

In the FAQ's section on the IDSA's website, when asked about treatment, it stated,

" When Lyme disease is diagnosed and treated quickly, 95 percent of people are cured within a few weeks of treatment.

The number of people who continue to have problems (the other 5%) is very small. Most likely, their symptoms are related to one of the following:

- They never had Lyme disease at all and received the wrong treatment for their illness
- They had Lyme disease and another infection simultaneously and were only treated for Lyme disease
- They contracted a new illness unrelated to Lyme disease but with similar symptoms
- They have again been bitten by the tick that causes Lyme disease." (What a vengeful tick- to come back and bite me twice!!)

The page then went on to comment on patients with lingering symptoms, typically pain and fatigue, must actually have some other disease or psychological condition because chronic Lyme is non-existent.

The IDSA believes that long term antibiotic therapy for " so called chronic Lyme" is not only unproven effective, but rather unsafe. I' m sorry, but the fear of C-diff which can properly be controlled with taking probiotics (acidophilus) to promote healthy growth of the intestinal flora in conjunction with the antibiotic will control this. Concerns about elevated liver enzymes can happen. Taking milk thistle capsules, the same ones given to cirrhosis patients and maintenance blood work, a procedure used in many other medications is the responsible way to alleviate this concern. What about their concerns about long term therapy creating super bugs? What is the IDSA thinking???? Isn' t that what they are doing by under treating the Lyme bacteria that then fights back in a more cystic and resilient form- hence creating Lyme super bugs? This is one of the reasons why the new infections of Lyme disease are far worse then the forms that were existent long ago. This occurrence is happening in many areas of medicine. To believe that they are not doing so is asinine.

The IDSA admitted to having reviewed over 400 reference papers to come up with these guidelines and many more were submitted did not reach their criteria. What they didn' t tell you was that those 400 reference papers were written by the doctors on that panel.

Attorney General Richard Blumenthal, announced that his antitrust investigation has *uncovered serious flaws* in the Infectious Diseases Society of America's (IDSA) process for writing its 2006 Lyme disease guidelines and the IDSA has agreed to reassess the guidelines with the assistance of an outside arbiter so the investigation of the IDSA would stop.

The IDSA guidelines have sweeping and significant impacts on Lyme disease medical care. They are commonly applied by insurance companies in restricting coverage for long-term antibiotic treatment or other medical care and also strongly influence physician treatment decisions.

Insurance companies have denied coverage and medications for long-term antibiotic treatments relying on these guidelines as justification. The guidelines are also widely cited for conclusions that chronic Lyme disease is nonexistent.

"This agreement vindicates my investigation −− finding undisclosed financial interests and forcing a reassessment of IDSA guidelines," Blumenthal said. "My office uncovered undisclosed financial interests held by several of the most powerful IDSA panelists. The IDSA's guideline panel improperly ignored or minimized consideration of alternative medical opinion and evidence regarding chronic Lyme disease, potentially raising serious questions about whether the recommendations reflected all relevant science.

"The IDSA's Lyme guideline process lacked important procedural safeguards requiring complete reevaluation of the 2006 Lyme disease guidelines −− in effect a comprehensive reassessment through a new panel. The new panel will accept and analyze all evidence, including divergent opinion. An independent neutral arbitrator −− expert in medical ethics and conflicts of interest, selected by both the IDSA and my office −− will assess the new panel for conflicts of interests and ensure its integrity."

Blumenthal's findings include the following:
- The IDSA failed to conduct a conflicts of

interest review for any of the panelists prior to their appointment to the 2006 Lyme disease guideline panel

- Subsequent disclosures demonstrate that several of the 2006 Lyme disease panelists had conflicts of interest such as 6 of the 14 members or the Universities in which they are employed hold patents for Lyme disease or their co-infections. 4 members received funding from test kit manufactures, 4 members were paid by insurance companies to write Lyme guidelines ideally allowing them to legally deny participants their treatment and medications, 9 members of the panel received funding from the authors or their Universities in which they are employed.

- The IDSA failed to follow its own procedures for appointing the 2006 panel chairman and members, enabling the chairman, who held a bias regarding the existence of chronic Lyme, to handpick a likeminded panel without scrutiny by or formal approval of the IDSA's oversight committee

- The IDSA's 2000 and 2006 Lyme disease panels refused to accept or meaningfully consider information regarding the existence of chronic Lyme disease, once removing a panelist (Dr. Sam Donta) from the 2000 panel who disagreed with the group's position on chronic Lyme disease to achieve "consensus".

- The IDSA blocked appointment of scientists and physicians with divergent views on chronic Lyme who sought to serve on the 2006 guidelines panel by informing them that the panel was fully staffed, even though it was later expanded.

- The IDSA portrayed another medical association's Lyme disease guidelines as corroborating its own when it knew that the two panels shared several authors, including the chairmen of both groups, and were working on guidelines at the same time. In allowing its panelists to serve on both groups at the same time, IDSA violated its own conflicts of interest policy.

Ultimately the members of the Infectious Disease Panels should have no monetary ties to create poor judgment and conflicts of interest when creating guidelines on how chronically ill patients will be treated.

IDSA has reached an agreement with Attorney General Blumenthal's office calling for the creation of a review panel to thoroughly scrutinize the 2006 Lyme disease guidelines and update or revise them if necessary. The panel -- comprised of individuals without conflicts of interest -- will comprehensively review medical and scientific evidence and hold a scientific hearing on July 30th 2009, to provide a forum for additional evidence. It will then determine whether each recommendation in the 2006 Lyme disease guidelines is justified by the evidence or needs revision or updating.

Blumenthal added, "The IDSA's 2006 Lyme disease guideline panel undercut its credibility by allowing individuals with financial interests -- in drug companies, Lyme disease diagnostic tests, patents and consulting arrangements with insurance companies -- to exclude divergent medical evidence and opinion. In today's healthcare system, clinical practice guidelines have tremendous influence on the marketing of medical services and products, insurance reimbursements and treatment decisions. As a result, medical societies that publish such guidelines have a legal and moral duty to use exacting safeguards and scientific standards.

"Our investigation was always about the IDSA's guidelines process – not the science" . The agreement with IDSA ensures that a new, conflicts–free panel will collect and review all pertinent information, reassess each recommendation and make necessary changes.

Under its agreement with the Attorney General's Office, the IDSA will create a review panel of eight to 12 members, none of whom served on the 2006 IDSA guideline panel. The IDSA must conduct an open application process and consider all applicants.

The agreement calls for the arbitrator selected by Blumenthal's office and the IDSA to ensure that the review panel and its chairperson are free of conflicts of interest. Blumenthal and IDSA agreed to appoint Dr.

Howard A. Brody as the arbitrator. Dr. Brody is a recognized expert and author on medical ethics and conflicts of interest and the director of the Institute for Medical Humanities at the University of Texas Medical Branch.

To assure that the review panel obtains divergent information, the panel will conduct an open scientific hearing at which it will hear scientific and medical presentations from interested parties. The agreement requires the hearing to be broadcast live to the public on the Internet via the IDSA's website. The Attorney General's Office, Dr. Brody and the review panel will together finalize the list of presenters at the hearing.

Once it has collected information from its review and open hearing, the panel will assess the information and determine whether the data and evidence supports each of the recommendations in the 2006 Lyme disease guidelines.

The panel will then vote on each recommendation in the IDSA's 2006 Lyme disease guidelines on whether it is supported by the scientific evidence. At least 75 percent of panel members must vote to sustain each recommendation or it will be revised.

Once the panel has acted on each recommendation, it will have three options: make no changes, modify the guidelines in part or replace them entirely.

The panel's final report will be published on the IDSA's website.

IDSA convened panels in 2000 and 2006 to research and publish guidelines for the diagnosis and treatment of Lyme disease. Blumenthal's office found that the IDSA disregarded a 2000 panel member (Dr. Sam Donta) who argued that chronic and persistent Lyme disease exists. The 2000 panel pressured the panelist to conform to the group consensus and removed him as an author when he refused. IDSA sought to portray a second set of Lyme disease guidelines issued by the American Academy of Neurology (AAN) as independently corroborating its findings. In fact, IDSA knew that the two panels shared key members, including the respective panel chairmen and were working on both sets of guidelines at the same time – *a violation of IDSA' s conflicts of interest policy.*

The resulting IDSA and AAN guidelines not only reached the same conclusions regarding the non-existence of chronic Lyme disease, their reasoning at times used strikingly similar language. Both entities, for example, dubbed symptoms persisting after treatment "Post-Lyme Syndrome" and defined it the same way. When IDSA learned of the improper links between its panel and the AAN's panel, instead of enforcing its conflict of interest policy, it aggressively sought the AAN's endorsement to "strengthen" its guidelines' impact. The AAN panel – particularly members who also served on the IDSA panel – worked equally hard to win AAN's backing of IDSA's conclusions.

The two entities sought to portray each other's guidelines as separate and independent when the facts call into question that contention. The IDSA subsequently cited AAN's supposed independent corroboration of its findings as part of its attempts to defeat federal legislation to create a Lyme disease advisory committee and state legislation supporting antibiotic therapy for chronic Lyme disease.

In a step that the British Medical Journal deemed "unusual," the IDSA included in its Lyme guidelines a statement calling them "voluntary" with "the ultimate determination of their application to be made by the physician in light of each patient's individual circumstances." In fact, United Healthcare, Health Net, Blue Cross of California, Kaiser Foundation Health Plan and other insurers have used the guidelines as justification to deny reimbursement for long-term antibiotic treatment.   Supplemental disability income insurance companies have also used these guidelines to reject payments on claims stating Lyme is not disabling.

Members of Blumenthal's office who worked on the investigation were Assistant Attorney General Thomas Ryan, former Assistant Attorney General Steven Rutstein and Paralegal Lorraine Measer under the direction of Assistant Attorney General Michael Cole, Chief of the Attorney General's Antitrust Department.[6]

So much respect and appreciation goes to Attorney Blumenthal and his investigating staff for the tremendous step they that allowed for the Lyme community. Without their help uncovering all of the deceit, lies and cover ups- we would not be in the position we are in now looking at the possibility within this year to have the treatment guidelines completely rewritten. If this review hearing goes well, this just very well may be the turning point that will cause a lot of people to get the treatments they need. This will allow protection to the doctors and patients involved in this illness. No longer will the insurance companies be allowed to deny our medications. Patients will no longer be forced to travel hours to see their Lyme specialists because more types of doctors will be able to be educated with this correct information. This was such an encouraging step and has really justified that something was truly wrong with how Lyme disease has been treated all along.

Chapter 4

The Keys to Unlocking Lyme

ॐ
Manganese
ॐ

To date, microbiologists at the University of Texas Southwestern Medical Center in Dallas, Texas have made a major breakthrough. These new findings may lead to new ways that we will be able to fight this infection.

A protein that is essential for the bacterium to become virulent (able to overcome bodily defensive mechanisms) has been discovered. This bacterial protein, called

BmtA, aids in transporting the metal, manganese, from the host tick or mammal to the Lyme disease causing bacterium Borrelia burgdorferi. Researchers studied Bb that was genetically engineered to lack the protein BmtA transporter and the effects that it had on mice. In a test tube the bacteria grew significantly slower but were not much more different from the original form. However, according to Dr, Michael Norgard, the chairman of microbiology at the University of Texas Southwestern, this genetically engineered bacterium when placed in a mouse could not grow.

It was also discovered that Lyme disease does not use iron to thrive. It is very rare that any type of bacteria does require this. Of the thousands of different types of bacteria out there, only a few species do not require iron to survive. Scientists are now researching what other minerals are taking the place of the iron's functions.

While these are major breakthroughs, researchers need to more exactly determine what other steps need to happen in order for Borrelia burgdorferi to become virulent. Manganese may be more of an indirect component, but essential regardless. Knowing this amazing discovery, we are much closer to finding new finding ways to stop the infections in patients and possibly finding new diagnostic tools.

❧✦❧

Biofilms

❧✦❧

Early in 2008, two respected scientists, Drs. Eva Sapi and Alan MacDonald were able to show that biofilms were part of the biology of Borrelia.

A biofilm is a complex structure of bacteria that functions like a community. It will secrete a slime like coating to protect itself and survive when it feels like it is being placed in a hostile environment defending itself from antibiotics and the immune system. In the cluster, the spiral form of the spirochete often mutates into cystic forms, granular dots, or L-forms. After the colony has built this protective slime shield, it is extremely resistant to antibiotics, allowing this infection to be unable to be killed off and it will continue to survive in the body.

The reason why this discovery is so important is that it can help show a theory on why some cases of Lyme are chronic and are not able to be killed off with short term antibiotics. This is why many doctors prescribe cyst buster medications along with the antibiotics.

There are some doctors that are presenting theories on using specific enzymes to break apart the biofilms. Biofilms are also present in the mouth in the form of plague. Using the model of studying what kills the spirochetes in the mouth; Scientists are looking at some of the items and essential oils in Listerine as one example on how to kill these biofilms. Additionally, biological chemists are also extracting a wide range of natural chemicals from various botanicals. Some are growing bacteria while others kill the bacteria but currently there is a wide range as to how little or how severe of damage it will cause on human membranes.

Biofilms have so far been discovered in European Lyme Borreliosis or Borrelia Afzelii, which is the skin manifestation Acrodermatitis chronica atrophicans (ACA). ACA is a condition where the skin gets very thin and very fragile, and it persists for many years, and is one of the late manifestations of Lyme disease in the skin. It occurs about thirty years after you've been infected. Studies that have been done in Europe have shown that colonies of Borrelia spirochetes are present in the skin. ACA contains colonies of Borrelia, and Drs. Sapi and MacDonald have shown that those groupings or colonies are actually Borrelia biofilms. These biofilms allow the bacteria to survive for long periods of time in the human body so they can reactivate many years later after someone has gone into remission or present itself later in life as other diseases such as ACA, MS, and Alzheimer' s disease. It is theorized, that if we can attack these biofims, we may be able to prevent certain diseases later in life such as Alzheimer' s or Parkinson' s and chronic forms of Lyme disease.

ও৽৽৶

Understanding Lyme Bands

ও৽৽৶

Many patients and health care workers are uneducated about how to interpret a Western Blot test. If a person has one "fingerprint band," they have Lyme disease. These Lyme specific bands are the 18, 23, 25,

31, 34, 39, 83 or 93. It does not matter what lab you have chosen, if one of these bands is positive even once—Lyme is present. IgeneX is by far the best lab to use. They only test for Tick Borne infections. Many patients are generally happy with MDL labs because they also test for more forms of Lyme then some of the other labs out there such as LabCorp or Quest.

In a Western blot, the testing laboratory looks for antibodies directed against a wide range of Bb proteins. If the patient has an antibody to a specific Bb protein, a "band" will form at a specific place on the immunoblot. For example, if a patient has antibody directed against outer surface protein A (OspA) of Bb, there will be a WB band at 31 kDa. By looking at the band pattern of patient's WB results, the lab can determine if the patient's immune response is specific for Bb. Many patients have noticed that their Western blot report usually contains two parts: IgM and IgG. These are immunoglobulins (antibody proteins) produced by the immune system to fight infection.

1. IgM is produced fairly early in the course of an infection, while
2. IgG response comes later.

Some patients might already have an IgM response at the time of the EM rash; IgG response, according to the traditional model, tends to start several weeks after infection and peak months or even years later. In some patients, the IgM response can remain elevated; in others it might decline, regardless of whether or not treatment is successful.

Similarly, IgG response can remain strong or decline with time, again regardless of treatment. Most WB results report separate IgM and IgG band patterns and the criteria for a positive result are different for the two immunoglobulins.

Finally, in setting up a nationwide standard for a positive WB, one makes several assumptions−

1. that all strains of Bb will provoke similar immune responses in all patients,
2. that all patients will mount a measurable immune response when exposed to Bb, and
3. that the IgG immune response will persist in an infected patient.

Unfortunately, none of these is always true.

Therefore, a judicious interpretation of Western blot results in a clinical setting should take into account both

1. the vagaries of the human immune response and
2. the possibility that strain variations in Bb might produce unusual banding patterns.

The CDC criteria for a positive WB are as follows:
    * For IgM, 2 of the following three bands: OspC (22-25), 39 and 41.
    * For IgG, 5 of the following ten bands: 18, OspC (22-25), 28, 30, 39, 41, 45, 58, 66 and 93.

How were these recommendations arrived at? The IgG criteria were taken pretty much unchanged from a 1993 paper by Dressler, Whalen, Reinhardt and Steere. In this study, the authors performed immunoblots on several dozen patients with well characterized Lyme disease and a strong antibody response and looked at the resulting blot patterns. By doing some fairly involved statistical analysis, they could determine which bands showed up most often and which best distinguished LD patients from control subjects who did not have LD.

They found that by requiring 5 of the 10 bands listed, they could make the results the most specific, in their view, without sacrificing too much sensitivity. ("Sensitivity" means the ability of the test to detect patients who have the disease, "specificity" means the ability of the test to exclude those who don't. Usually, an increase in one of these measures means a decrease in the other.) The IgM (newer infection) criteria were determined in much the same fashion. Fewer bands are required here because the immune response is less mature at this point. Several studies have shown that

1. the first band to show up on a Lyme disease patient's IgM blot is usually the one at 41 kDa,
2. followed by the OspC band (22-25) and/or the one at 39.

The OspC and 39 kDa band are highly specific for Bb, while the 41 kDa band isn't. That's why the 41 by itself isn't considered adequate. Here's the rub, though: the CDC doesn't want the IgM criteria being used for any patient that has been sick for more than about six weeks. The thinking here is that by this time an IgG response should have kicked in and the IgM criteria, because they require fewer bands, are not appropriate for patients with later disease.[2] Another large complication is that the protein bands can "break apart" over time and not be present month to month so to allow the perfect combination of CDC criteria usually comes with perfect timing.

Chapter 5

Lyme Conspiracies and Theories- Fact or Fiction

❧

Biological Experimentation

Plum Island has always been a favorite controversy in the Lyme community. Even this topic will divide the effected. Plum Island is off the northeastern tip of Long Island, NY. It houses the Plum Island Animal Disease Center.

There is a high possibility that Lyme Disease may have spread as a result of experimentation for biological warfare research

on Plum Island—a Department of Agriculture
facility (recently in 2003 was joined by
Homeland Security taking responsibility for
safety and security of the facility) that doubled
as an Army biological warfare research facility.
Dedicated to the study of animal diseases, Plum
Island appears to have been the site of
experiments with disease-infected ticks
conducted by Nazi scientists brought into the
United States under " Project Paperclip" in
the 1950' s. One of the Nazi scientists who
appeared to have been involved with Plum
Island was Dr. Erich Traub, who was in charge
of the Third Reich' s virological and
bacteriological warfare program falsely named
a Nazi " Cancer Research Program" in World
War II. Was Traub involved with experiments
that led to the spread of Lyme disease?

  " In the preface of The Belarus Secret,
John J. Loftus laid out a striking piece of
information gleaned from his spy network:
' Even more disturbing are the records of the
Nazi germ warfare scientists who came to
America. They experimented with poison ticks
dropped from planes to spread rare diseases. I
have received some information suggesting that
the U.S. tested some of these poison ticks on
the Plum Island artillery range off the coast of

Connecticut during the early 1950' s. Most of the germ warfare records have been shredded, but there is a top secret U.S. document confirming that ' clandestine attacks on crops and animals' took place at this time."

There have also been 2 files titled " Tick Research" and " E. Traub" that were emptied but the folders remained. Ticks that were native to Texas suddenly appeared on the East Coast. Questions remained unanswered if during experimentation the ticks didn' t cross the water to the mainland of Connecticut by hitching a ride on the birds that would travel back and forth. Or even worse to imagine, did the people of the United States become the guinea pigs again for government testing??

Author Michael Carroll, " Lab 257: the Disturbing Story of the Government' s Secret Plum Island Germ Laboratory" , covers this article in great lengths for readers looking for the whole story.

The reason people have a right to question what may be happening relates back to the Tuskegee Syphilis Experiment. It feels all too familiar to the Lyme disease controversy.

' For forty years between 1932 and 1972, the U.S. Public Health Service (PHS) conducted an experiment on 399 black men in the late stages of syphilis. These men, for the most part illiterate sharecroppers from one of the poorest counties in Alabama, were never told what disease they were suffering from or of its seriousness.

Informed that they were being treated for " bad blood," their doctors had no intention of curing them of syphilis at all. The data for the experiment was to be collected from autopsies of the men, and they were thus deliberately left to degenerate under the ravages of tertiary syphilis—which can include tumors, heart disease, paralysis, blindness, insanity, and death. " As I see it," one of the doctors involved explained, " we have no further interest in these patients until they die."

The true nature of the experiment had to be kept from the subjects to ensure their cooperation. The sharecroppers' grossly disadvantaged lot in life made them easy to manipulate. Pleased at the prospect of free medical care—almost none of them had ever seen a doctor before—these unsophisticated and trusting men became the pawns in what

James Jones, author of the excellent history on the subject, " Bad Blood" , identified as " the longest non therapeutic experiment on human beings in medical history."

The study was meant to discover how syphilis affected blacks as opposed to whites— the theory being that whites experienced more neurological complications from syphilis whereas blacks were more susceptible to cardiovascular damage. How this knowledge would have changed clinical treatment of syphilis is uncertain. Although the PHS touted the study as one of great scientific merit, from the outset its actual benefits were hazy. It took almost forty years before someone involved in the study took a hard and honest look at the end results, reporting that " nothing learned will prevent, find, or cure a single case of infectious syphilis or bring us closer to our basic mission of controlling venereal disease in the United States." When the experiment was brought to the attention of the media in 1972, news anchor Harry Reasoner described it as an experiment that " used human beings as laboratory animals in a long and inefficient study of how long it takes syphilis to kill someone."

By the end of the experiment, 28 of the men had died directly of syphilis, 100 were dead of related complications, 40 of their wives had been infected, and 19 of their children had been born with congenital syphilis. How had these men been induced to endure a fatal disease in the name of science? To persuade the community to support the experiment, one of the original doctors admitted it " was necessary to carry on this study under the guise of a demonstration and provide treatment." At first, the men were prescribed the syphilis remedies of the day—bismuth, neoarsphenamine, and mercury—but in such small amounts that only 3 percent showed any improvement. These token doses of medicine were good public relations and did not interfere with the true aims of the study. Eventually, all syphilis treatment was replaced with " pink medicine" —aspirin. To ensure that the men would show up for a painful and potentially dangerous spinal tap, the PHS doctors misled them with a letter full of promotional hype: " Last Chance for Special Free Treatment." The fact that autopsies would eventually be required was also concealed. Even the Surgeon General of the United States participated in enticing the men to remain in the experiment, sending them certificates of appreciation after

25 years in the study.

A Tuskegee nurse explained her role as one of passive obedience: " we were taught that we never diagnosed, we never prescribed; we followed the doctor's instructions!"

One of the most chilling aspects of the experiment was how zealously the PHS (Public Health Service) kept these men from receiving treatment. When several nationwide campaigns to eradicate venereal disease came to Macon County, the men were prevented from participating. Even when penicillin was discovered in the 1940s—the first real cure for syphilis—the Tuskegee men were deliberately denied the medication. During World War II, 250 of the men registered for the draft and were consequently ordered to get treatment for syphilis, only to have the PHS exempt them.

Pleased at their success, the PHS representative announced: " So far, we are keeping the known positive patients from getting treatment." The experiment continued in spite of the Henderson Act (1943), a public health law requiring testing and treatment for venereal disease, and in spite of the World Health Organization's Declaration of Helsinki

(1964), which specified that " informed consent" was needed for experiment involving human beings.

The story finally broke in the Washington Star on July 25, 1972, in an article by Jean Heller of the Associated Press. Her source was Peter Buxtun, a former PHS venereal disease interviewer and one of the few whistle blowers over the years. The PHS, however, remained unrepentant, claiming the men had been " volunteers" and " were always happy to see the doctors," and an Alabama state health officer who had been involved claimed " somebody is trying to make a mountain out of a molehill."

Under the glare of publicity, the government ended their experiment, and for the first time provided the men with effective medical treatment for syphilis. Fred Gray, a lawyer who had previously defended Rosa Parks and Martin Luther King, filed a class action suit that provided a $10 million out-of-court settlement for the men and their families.

The PHS did not accept the media's comparison of Tuskegee with the appalling experiments performed by Nazi doctors on their Jewish victims during World War II. Yet in addition to the medical and racist parallels, the PHS offered the same morally bankrupt defense offered at the Nuremberg trials: they claimed they were just carrying out orders, mere cogs in the wheel of the PHS bureaucracy, exempt from personal responsibility.

The study's other justification—for the greater good of science—is equally spurious. Scientific protocol had been shoddy from the start. Since the men had in fact received some medication for syphilis in the beginning of the study, however inadequate, it thereby corrupted the outcome of a study of " untreated syphilis." [4]

As preposterous and paranoid as this Plum Island theory may sound, at one time the Tuskegee experiment must have seemed equally farfetched. Who could imagine the government, all the way up to the Surgeon General of the United States, deliberately allowing a group of its citizens to die from a terrible disease for the sake of an ill-conceived experiment?

Another inhumane biowarfare military experiment came in the form of " Project Eight

Ball" . Housed in Fort Detrick MD, the home of the United States Army Medical Research Institute of Infectious Diseases, is a 40-foot-high stainless steel sphere. In a secret operation carried out under Project Eight Ball, was Operation Whitecoat, which titled the testing' s during 1954-1973 of over 2300 soldiers. Most soldiers, eager to serve their country, were seated on a cat walk outside the middle of the sphere that everyone on base called the "Eight Ball." Soldiers wore a rubber mask connected to a breathing tube that brings in air from inside the sphere. Industrial fans kick in, and within a few minutes, the experiment is over.

A few days later the fever, cough and aches set in. Participants were being infected with Q fever, Yellow fever, plaque, Tularemia, Hepatitis A, Venezuelan equine encephalitis, and many other serious diseases.

Soldiers were then treated for their inflicted illnesses to test the effectiveness of experimental medications and vaccines. Most soldiers were only given 2 weeks of leave as reward for their participation. Others did it to fulfill their obligations to the military due to religious reasons allowing them to not engage in combat. Much of the testing remains classified and Fort Detrick allows no visitors. Not even ex-soldiers who were exposed as part of the experiment.

In 1969, President Nixon signed an order ending offensive biological weapons programs, signaling the beginning of the end of Operation Whitecoat, which closed for good in 1973. Since that time any research done at Fort Detrick has been purely defensive in nature, focusing on diagnostics, preventives and treatments for biowarfare infections. However the military has found new ways to continue their vaccine testing on subjects. Currently, the enlistees are lined up for undisclosed shots. They ask you what your name is, have you stand on a set of yellow foot prints and then administer the injections. Then you were given the rest of the day off. Upon review of their military records, enlistees cannot find record of when these shots were administered anywhere in their paperwork or find they will bulk it under something cryptic to the effect of 6 tetanus shots or extra flu vaccines. When adverse reactions would occur, the infected would be told that it is confidential or classified information when inquiring what they were injected with.

There are a rising number of US service men and women that are becoming sick after receiving mystery vaccines. There are many documented cases of military personnel reporting they were having reactions yet these were never reported to the CDC and FDA or listed in their medical files upon review. There has even been a case where a hand written note was inserted into the medical records stating this person had a flu vaccine after TV media started investigating one case.

Many victims have been turned away being told they were nuts and their symptoms are all in their head. It is believed that the Department of Defense yet again is using their service men as guinea pigs testing vaccines for AIDS, Lyme disease, etc. 80% of service men forced to take the 6 shot Anthrax Vaccine suffer side effects according to a 2002 General Accountability Office report. Soldiers that would outright refuse inoculations are faced with military punishment such as a favorite Field Issue Article 15 which entails confinement till court hearing and/ or monetary fines. Many have testified that they were left to have their superiors punish their whole platoon to try to turn the unit against them or try to threaten them that if they did not take the shot voluntarily, it will be forced on them in their sleep- without a darn thing they could do about it. Some even have been assigned extra duties, taken off of missions, dropped in military rank and pay scale.

The connection with Lyme disease, it is feared that the Department of Defense has been testing new Lyme disease vaccines on the service men since the initial vaccine was pulled from the market. Many people who came out of the service, diagnosed with Gulf War Syndrome for example; were later diagnosed with Lyme disease.

While I am not taking a side in whether this is happening or not, there is plenty of evidence out there showing a lot of wrong doing in the past and a lot of cover up going on in the present.

ॐॐ
Transmission

New modes of transmission have always been discussed. It has been proven that many species of ticks, not just deer ticks can pass

Lyme. It also has been questioned weather a mosquito, flea or other blood sucking insect could pass this bacterium. The bacterium has been already been proven present in mosquitos, fleas, mites, and well water. It is quite plausible since it is a vector borne pathogen and the cycle does fit. The only question has been the timeline needed for the infection to invade the body. Theories have been presented about where a bite should occur, is to how fast you can be infected example being right over a vein.

Lida Mattman, who is a microbiologist and the author of "Stealth Pathogens", has studied spirochetes for fifty years. She believes that even touching can spread Lyme disease. The spirochete is found in tears, which means that it can contaminate the hands and anything they touch. Scientists are finding that the Lyme spirochete is very hardy and can remain viable for long periods of time. Spirochetes are also found in breast milk and semen. Could this possibly be a mode of transmission for Bb within families? It is not uncommon for whole families to be infected with Lyme disease or even become chronically ill. The CDC' s opinion from prior CDC Director, Julie Louise Gerberding, MD (who resigned from her position at the request of President Obama), is that families only can become infected by living in epidemic areas and they must not be taking precautions. What about all the other families in that epidemic area that are not taking precautions as well, but not one member of their family is sick, yet my neighbor and all of her family is? Where is the science in that?

When a pregnant woman is infected with Lyme disease, not only is she subject to its devastation, but her baby is too. There are a large number of mothers that feel that their children had received Lyme disease congenitally while they were pregnant. The IDSA and especially Eugene Shapiro, feels that this is ludicrous. That there is no way that this can happen. However, if you study the model of Syphilis which is also a spirochete just like Lyme, you can plainly see that spirochetes very much can be passed via vertical transmission, mother to placenta. It is insane to not realize this.

The first case of transplacental passage of Borrelia burgdorferi was reported in 1985 in Wisconsin by Schlesinger. The woman was bitten during her first trimester and developed an EM rash with two satellite lesions. This was followed by typical Lyme symptoms. She did not receive medical treatment as Lyme was not diagnosed at the time. She delivered a male baby at 35 weeks. The baby died 39 hours later from congestive heart failure, and at autopsy there were several major defects of the heart. Spirochetes were found in the spleen, kidneys, bone marrow and the heart. The mother tested positive for Lyme. Here we can only speculate that the Lyme might have been responsible for the birth defects, as these same types of problems can occur in non-Lyme situations.

In 1987, Dr. Alan MacDonald reported a case of a woman infected with Lyme in her first trimester of pregnancy, which unfortunately was not diagnosed or treated. She had developed a circular red rash which was followed by painful swelling of her knee. These resolved spontaneously. The woman went into labor at term, and delivered a 2,500 gram stillborn baby. Autopsy revealed a ventricular septal defect, i.e. a hole in the wall of the heart which separates the two main pumping chambers. The Lyme bacterium was cultured from the baby' s liver, and it was demonstrated in the brain, heart, adrenal gland and the placenta. The mother' s blood tested positive for antibodies to the Lyme spirochete and negative for syphilis. Dr. MacDonald reported three other cases of fetal death in the second trimester, in which the Lyme spirochete was cultured from the livers. None of the mothers gave any history suggesting Lyme infection.

In 1986, Weber reported a case of Lyme infection in a newborn baby. The mother had been bitten by multiple ticks during her first trimester. She developed an EM rash several weeks later. She was treated with a " standard" course of oral penicillin for seven days, three times a day. The baby was delivered at term and appeared normal. During the next 23 hours the baby developed breathing problems and died. Autopsy showed brain hemorrhages. Spirochetes compatible with Borrelia burgdorferi, the Lyme spirochete, were demonstrated in the brain and the liver. Initial testing of the mother' s blood was negative for antibodies to the Lyme spirochete; however, at a later date her frozen blood tested positive for IgM antibodies by the ELISA test.

Markowitz published a study of Lyme and pregnancy in 1986. He described nineteen patients who were infected during pregnancy.

Five of these had adverse outcomes (one fetal death at 20 weeks, high bilirubin level in a four-week premature baby, webbed toes, blindness and developmental delay, and a newborn rash). Thirteen of the nineteen had received antibiotics. The authors concluded that there was no proof that Lyme was responsible for the adverse outcomes since all of them were dissimilar. However, there was a consensus that this was an abnormally high frequency of adverse outcomes, and that pregnant women with diagnosed Lyme should be treated immediately with penicillin.

Williams and colleagues conducted a study in a Lyme-endemic area in New York of umbilical cord blood. Of 255 infants tested, 10.2% had detectable antibody to the Lyme spirochete. Of 166 infants born in a non-endemic area, 2.4% had detectable antibodies. The rate of birth defects did not differ significantly between the two groups; however, the first group tended to be of lower birth weight and smaller for their gestational age, and tended to have more jaundice. The authors concluded that these differences were not significantly different. A glaring flaw in this study is that it only included live births. Since miscarriages, stillbirth and perinatal infant deaths were not included, the possibility of congenital defects possibly associated with Lyme and incompatible with life are not included. Therefore, the author's contention that no association exists between gestational Lyme and congenital defects should be viewed with skepticism.

Dr. Andrea Dlesk, of the Marshfield clinic in Wisconsin, studied 143 healthy pregnant women. Lyme serologic tests were obtained on the initial and postpartum visits. At the time the data were reported, 116 women had completed their pregnancies and 12 had miscarried, one of whom tested positive. Of the 104 women who did not miscarry, 13 women tested positive for Lyme. The conclusion was that healthy women who test positive for Lyme are at no increased risk for miscarriage. Again this study is flawed in that there are no autopsy data on the 12 miscarriages. It is quite possible that, in the 11 seronegative mothers who miscarried, seronegative Lyme was present and may have caused defective fetuses. Seronegative Lyme is a real entity and may account for 25% of all cases of Lyme.

In 1989, Dr. Alan MacDonald reported his findings in autopsies performed following perinatal deaths at Southhampton Hospital between 1978 and 1988. It must be noted that routine pathology studies on tissues will not demonstrate the Lyme spirochete. Unless there is a high index of suspicion for Lyme disease, the special silver or immunologic stains which can identify the spirochete are not used. He also reports four cases where there was live birth and spirochetes were demonstrated in the placentas. In the group of perinatal deaths there was no history or evidence of Lyme disease in the mothers. Their blood tests were negative in all but one case. Spirochetes compatible with Borrelia burgdorferi were identified in the vital organs and numerous developmental defects were observed. Dr. MacDonald' s conclusions are:

1.   Tissue inflammation is not seen in fetuses with transplacentally acquired infection with the Lyme spirochete.

2.   Lyme disease acquired in utero may result in fetal death in utero, fetal death at term or infant death after birth. Babies may also survive in spite of the bacteria being isolated in the placenta.

3.   In all but one of these cases where the

Lyme organism was identified in the placenta or the fetal tissues, the maternal blood had no evidence of antibodies to the Lyme bacteria. In only two of the fourteen cases was there a maternal history compatible with Lyme disease, yet neither of the two were serologically confirmed.

This is the extent of the currently available information on Lyme and Pregnancy in the medical literature in 1990. Comparing the various studies has led us to arrive at the following conclusions:

1.   Lyme disease is a serious threat to pregnant women in that it may cause fetal damage and death.

2.   Pregnancy may mask symptoms of Lyme in the mother and may result in seronegativity.

3.   Serologic screening of pregnant women in highly endemic areas is not recommended.

4.   Pregnant women who test positive for Lyme antibodies, yet have no symptoms suggesting active Lyme, are probably at a lower risk of passing the infection across the placenta. It may be possible that the presence

of antibody prevents the Borrelia burgdorferi from crossing the placenta.

5.   Babies born with Lyme disease can be expected to have a negative blood test for Lyme antibodies. Few have positive test.

6.   We desperately need a better test for detecting Lyme in pregnant women. It is clear that serologies are inadequate. Efforts should be directed at evaluating urine antigen and PCR testing in pregnancy and in neonates.[5]

While the CDC does not confirm those studies, it does state on its Web site: "Lyme disease acquired during pregnancy may lead to infection of the placenta and possible stillbirth."

If ticks were the only vector for Lyme disease, the American Red Cross, the Food and Drug Administration and the Centers for Disease Control and Prevention would not stop Lyme disease patients from donating blood. In their guidelines, it states: "Lyme disease: If this is a chronic condition you cannot donate. If you were treated with antibiotics and completely recovered, you can donate 12 months after the last dose of antibiotics was taken." The American Red Cross also states that anyone who has ever had Babesia can never donate blood. Babesia is another tick-borne disease, often a co-infection with Lyme. Same is true for organ donors, your organs will not be accepted if you have Lyme disease.

ॐ
LYMErix

The Lyme vaccine, LYMErix, was pulled from the market in 2002. GlaxoSmithKline said it was due to low sales. There are a large percentage of Lyme suffers who claim they contracted Lyme disease from this vaccine. In fact, the CDC stated that it activated a Lyme band in the Western Blot testing and therefore it was no longer allowed as part of the diagnostic criteria. By eliminating a whole band of proteins in the testing, you are eliminating a whole group of people who are infected by that same strain they based LYMErix off of. This vaccine was based off of the Lyme Protein OspA. This protein is seen as Band 31 on a Western Blot test. Not surprisingly, some people have accused the OspA vaccines of having caused severe knee damage by stimulating destructive arthritis. This correlates well with the fact that Lyme arthritis almost invariably affects the knees but other joints with less frequency. This pattern reflects the tremendous physical stress imposed on the knees inherent to knee kinesiology. OspA is considered an "arthrogenic antigen".

To date, SmithKline, IDSA' s Gary Wormser and Robert Nadelman were all sued numerous times for exacerbating chronic Lyme and Adverse Lyme effects. Wormser was also found to have known before hand that OspA blunted the immune response in vaccinated animals in a published medical report from July 28th 2000. However, GlaxoSmithKline and the CDC claim that not one person received Lyme disease from this vaccine.

Chapter 6

Would You Walk A Mile In Our Shoes???

Extraordinary people survive
under the most terrible
circumstances and they become more
extraordinary because of it.
-Robertson Davies

Ahead are the stories of many people living with Lyme disease and in their own words through the journey that it has taken them.

Once you are diagnosed with Lyme you are part of a group of people, that are some of the strongest and most courageous and caring people in the world. Most of us have a lot to live for and lost a lot when this disease changed our lives. It is easy to see that we were not lazy or depressed individuals looking for attention when we were giving up fantastic careers, sacrificing our dreams or losing years of our children' s lives.

One of the hardest challenges in Lyme is that it is such an internal disease. People will often tell you," Well, you don' t look sick at all." Often you will hear from us, that it is one of the comments that hurts us the most. If others only understood the war that was raging under our skin, they might understand that you don' t have to look like death to feel like it. As a chronic condition, we take great strides to live as normal of a life as possible. The only way to truly know, is to experience it for yourself. Please come see the world as we have been made to experience it.

Tara Hulko
Tamaqua, Pennsylvania

*"One of the most sublime experiences we can ever have-
is to wake up feeling healthy after we have been sick".*
-Harold Kushner

It is difficult to tell you where my journey began. I do remember, however, when I could no longer function. Like most people, I did not see the tick that bit me.

My family and I bought our first home and relocated to Schuylkill County, Pennsylvania in May 2007. We were so happy to make the transition. We found an old Victorian era mansion that needed a small bit of TLC. My husband had a good paying job that was worth the commute from our old home and I just starting into banking. Showing a lot of promise, I was cross training to become a financial consultant with a major institution. Our children loved their new home and for us, life couldn' t be better.

Then the 1st tragedy struck. My husband started severe migraine headaches right after we had moved. Progressively each month, he was getting worse and worse. New symptoms began to arise and he lost his job due to attending too many doctor appointments. By September, I arrived home from work to find him unable to walk, speak, and he was in excruciating pain. My husband was rushed to the hospital and admitted for multiple days for observation. It seemed his nervous system was shutting down and the physicians were unsure if he was going to pull through. I feared losing him and raising our children by myself. In the end, the doctors could not find anything significantly wrong and said it must have just been an extreme case of migraine headaches and sent him home. After that day, he was unable to use his right leg and has a severe pain disorder, no reflexes and neurological problems. Effectively, I was left having to care for my husband, my children and still work full time at the bank. Surprisingly, I was doing it and still excelling at my career.

October came and I was hired to run a haunted hayride for my boss at the bank. It was a lot of fun, because she had a very large amount of land back in the woods close to Hawk Mountain. My husband helped me organized an amazing haunted woods in spite of the pain he was in, we thought it might be entertaining and it would help take his mind off of things. The actors and I sat back in the woods attempting to scare kids for many hours all through the night.

Then in late October, I started feeling tired, lightheaded and having daily headaches at work. I really thought nothing of it considering the amount of stress I was under. I was managing by large amounts of coffee and over the counter pain medication. This trend continued for quite some time.

On December 17, 2007, I took a turn for the worst. I was at work and I started to feel rather ill. I felt like I was going to pass out, I was weak and trembling. I remember feeling very confused, my heart rate was sky high and everything felt surreal. I left work and went to the ER at a local hospital. My resting heart rate was up in the 170' s. It was explained that I must be over stressed and was sent home to relax.

A few days later, I was admitted for the same thing. This time my heart rate needed to be brought down with a beta blocker heart medication. This doctor ran more tests, and said he felt I was suffering a pre-heart attack. I was 27 years old and suffering heart problems?!? So I was given a prescription and referred to a cardiologist.

Ongoing, both my family doctor and cardiologist felt it must just be stress. I was started on antidepressants but also continued with the heart medication, just in case they were wrong. Neither medication helped. And the testing continued.

Over the next few months I was back in the ER a total of 8 times including my birthday and Valentines Day. My resting heart rate was over 200 at one point (while my normal is 80) plus it was skipping beats. I was also dealing with hard shaking. I felt that I looked like someone having drug withdraw symptoms. I was afraid because I was young, that the doctors would think that was my problem. Then I would wait for them to draw blood and when the labs came back, they saw I was clean. At that point the doctors would treat me like someone who actually had something going on. In addition, I did get the honor of wearing a heart monitor for 21 days which during that time, I did not have any major attacks.

Around February 2008, I started to get even worse. I started having horrific pain in the back of my neck, severe fatigue, became so dizzy and light headed I was walking into walls, suffering neurological complications such as watching walls and objects melt and distort in front of me, constantly trembling, sometimes convulsing, and I was in searing full body pain. During one of my ER visits, I asked the doctor if I could possibly be having seizures. He sent me to a neurologist. The neurologist confirmed that I had abnormalities after EEG testing. I was then placed on seizure medication. I also suffered a 4 day in hospital study of my seizures to then be told that I am creating these due to stress. I was also told that I am manifesting these seizures as a coping mechanism of something traumatic that must have happened to me as a child like a rape or molestation. Even though these things never happened, I was told I should look into options like psychological therapy.

I just wasn' t making any headway and kept getting worse. I was given so much testing and no one found any answers. The doctor' s were starting to not know where to turn. Eventually in May 2008, there was a bull' s eye rash on my leg. Now, growing up in New Jersey, where they educate on Lyme Disease at a young age, I knew immediately, this was a possibility. I ran to my computer and looked to see what else could cause a rash like that. Ringworm. So I took a picture of my rash and made an appointment with my doctor.

Upon going to my family doctor, she said this is definitely a bull' s eye caused by a tick and started me on the oral antibiotic, Doxycycline, till we could get my test results back. She told me 21 days would be more then enough to knock it right out of me.

Well, my tests came back and I was told I was negative for Lyme. I told her I knew this Elisa test to be unreliable and she agreed to continue my treatment and retest again after 3 months. 3 months came and went and I got slightly better, but not well enough to return to work yet. I was sent for my Lyme retest and after complaining about my debilitating fatigue, she ran a mono test as well. I was sleeping about 12 hours a day and still exhausted.

My testing came back negative again for Lyme, but I did have an old mono infection and low vitamin D. I figure when I was dealing with the worst of it in February, I was also fighting Mono as well. I did ask to have a Western Blot test done so I can see if I have any reactive bands. I was told by my family doctor to " get this idea of Lyme Disease out of my head" and " I really am dealing with depression" . I told her I knew I wasn' t depressed. I knew there was more going on. I asked her " how did I get a bull' s eye rash then" and she said, " well spiders can do that too" ! (Side note: No, they cannot. I asked my Lyme Literate Doctor. The only way to get this bull' s eye I had was through Lyme because it was not ringworm).

I was really sick of hearing that it is all in

my head and that stress is why I am this ill. It was the catch all diagnosis. If you couldn' t find an answer, it must be stress. I told her someone really needed to look at the big picture and stop treating one symptom at a time. Eventually she did agree to send me to a Rheumatologist for the pain and fatigue.

I researched Lyme Disease through and through and weeded out the good and bad information because there are a lot of both out there. I decided enough was enough and contacted a Lyme Literate Doctor near Philadelphia, Pa. He was over 2 hours from my home, but no one around here treated for Lyme let alone took insurance too. He diagnosed me with Lyme Disease after running numerous tests that my family doctor would not administer and he has continued my treatment. My Rheumatologist also diagnosed me with Fibromyalgia and Chronic Fatigue Syndrome brought on by the Lyme Disease. These are not diagnoses that replace Lyme, but rather help to prove that Lyme had affected my body permanently. I also was diagnosed with sinus tachycardia from a new cardiologist.

My typical days involve seizures, severe fatigue, cognitive problems to which I cannot do simple math or spell correctly, incorrect word association, daily headaches or migraines, pain in my joints and directly into my bones, burning throughout my muscles, heart palpitations and irregular beats, occasional blurred vision, vertigo, shortness of breath, loss of memory and confusion, pain in my meninges, weakness, tremors, nausea, lightheadedness, and oversensitivity to my senses. I cannot use a computer for any length of time or read due to the fact it exacerbates out the severe fatigue. I am also gaining a fear of going out without a family member because I am afraid of seizing. I was already trapped somewhere for 5 hours once because the seizures came on and off and I couldn't drive home in my condition. These aren't all of my complications I have come up against, however these are the ones I usually deal with daily in one extreme to the next.

During the course of this, I was continuing to fight with my employer's disability provider because they didn't feel that my Lyme could have lasted more then 2 weeks and the fact that for awhile I didn't have any tests proving I even had Lyme according to the CDC's standards allowed them a loop hole. I would go weeks and also at one point, 4 months, with no income in our household. My husband was still awaiting Social Security, so he had no income as well.

There aren' t many programs that will help you in the short term and long waiting lists for the ones that really can help you. We had utilize a food bank while we awaited food stamps. We became a burden on our families so we would not lose our home. We had to heat our home in the worst of winter by turning on our oven, opening the door and letting that supplement the warmth because we could not afford oil and we didn' t have any additional fuel assistance.

Our savings are gone. We were about to open up our own business, an upscale café in our small town. Unfortunately, I don' t know if that can happen, but I still hope that it will. I was a professional model working with some famous names. I tried modeling after the infection and I became very injured on a shoot. All of my skills involve utilizing a computer and with a large part of that being taken away, I am left with not much I can do for employment. I am too confused to take orders, too weak and sore for repetitive motion and too exhausted to be awake and active more then 2-3 hours.

Ultimately Lyme has caused our family to suffer quite a bit. Sometimes people don' t realize that we don' t want to be sick, we don' t want to give up all that is good in life, yet they think we are making all this up for attention. Some believe that we choose to be " lazy" and give up our dreams and comfortable lifestyles. Why would anyone willingly do that?

Upon watching my husband' s illness, I was certain my husband had Lyme as well, because he remembered two bulls eye rashes on his body while in the Army. He was never tested then and therefore left untreated. He shares all my same neurological symptoms and quite a few others. He was also diagnosed with Fibromyalgia and unexplainable ailments. I had sent him to my Lyme doctor and with 1 Western Blot test, he had 3 Lyme specific bands pop up positive and was a few points shy of another 1 being positive as well.

For me, I am uncertain if I became infected through a tick bite or through my husband. It is hard to say. There is research out there that supports Lyme can be sexually transmitted like syphilis but no one is certain. Both bacterium are spirochetes and are very similar in their behaviors. I do find it ironic that we both became ill with extremely similar symptoms a few months apart from each other.

I do know, however, that in my case, I am infected with a very strong and resilient strain of Lyme due to the fast and quickly debilitating reaction in my body and also due to its intelligent hiding nature. I was lucky to be someone who had the EM bulls eye rash, because if I hadn' t (even though I saw it later in my infection, it can reoccur in some people), we would still be searching for something else. It was the only clue I had to lead me to a proper diagnosis. Unfortunately, I think I went 8 months undiagnosed with this strong disseminated strain, so that is why I am slow to get better.

It is so hard to forget about being ill when you are this sick. You feel so terrible and in my case, for example, I have had to schedule in 26 pills a day just so I can function. It is very hard to work them all in since some of them cannot be taken close to others and some even come with food restrictions. I even have had to keep a chart so I would not forget what I have taken due to my cognitive difficulties and loss of short term memory problems. I always feared overdose by taking a double dose.

I miss being able to be the mother my children once had. While I do spend as much time as I can with them, finding new ways to play with them, it is hard trying to take them to the park let alone something like bowling or roller skating. I am a mother of 2. I have a 7 year old son and a 3 year old daughter. I am also a step mom to 2 other daughters, ages 10 and 14. Of all the children, I think my son has taken it the hardest. He watches his mommy so ill and often whisked away to the ER. He has watched me lay in bed or on a couch hooked up to numerous monitors for up to 3 weeks at a time. He has watched his mommy go through seizures many times. No child should have to be put through that and be made to understand illness. The worry and stress that they have to go through is unfair to them. No amount of reassurance can cover up true illness. The have a hard time understanding why the doctor cannot make you better.

Hopefully, we can help educate the public and bring this to the forefront. Like cancer and AIDS, awareness had to start somewhere. We need more people to fundraise and rally on their good days because if we don' t, we may never win against this chronic disease and many more will go untreated because they were misdiagnosed by doctors who are uneducated in this field. I personally hope in writing this book that it can make a difference, even if it's small.

Update: October 2008.
For the first time I am starting to see progress. My Lyme doctor and I changed my medication to Ceftin and I have also started a few supplements I found on my own. For the 1st time, I am starting to feel better. I am by no means cured or 100%, but my symptoms are lessening. I am so happy I get the opportunity to share that because in the long course of writing this book, I was starting to think I was dealing with the permanent effects of Lyme and I was no longer an active case. I'm killing off this bacteria!!! I feel I am going to win in the end. Hopefully I will only suffer minor lasting effects in the end. But I finally feel hope!!!!!!

Update: April 2009
I have come to the conclusion that my road ahead will be very long in this illness. My liver had started failing back in January and I had to be taken off of my Lyme medication all together. Luckily, my condition has not gotten

too much worse, but I am definitely not getting any better. It is easy to see the permanent effects Lyme has left me with. I can only hope my liver will heal enough to go back on the medicine.

In addition— my insurance company has denied my IV treatment and also my medication for my Fibromyalgia in spite my doctors submitting additional paper work stating it is all medically necessary. It is due to the fact of how expensive my medications are. It amazes me that someone who is not a doctor can decide what medicines I should be taking.

None the less, I am still standing strong. I am 29 years old and I look forward to a long life ahead of me. I have finished this book over the last year and I am currently appealing a decision for Social Security Disability, since they feel Lyme disease is not a chronic illness. Hopefully people will wake up and realize how debilitating this way of life is. My love and best to anyone fighting a chronic illness. I always hope no one else will ever have to really walk a mile in our shoes.

Publishers Update: The week before the publication of this book, we received a call alerting us that Tara

was admitted into the hospital and was placed in the Intensive Care Unit. She suffered cardiac complications and was diagnosed with Paroxysmal Supraventricular Tachycardia (PSVT) and a minor heart attack. We wish her the best and pray for her to have swift recovery.

Lauren
Las Vegas, Nevada

It's hard to tell my Lyme story from the beginning, since I don't know when the beginning really is. You see, I don't know when I first got Lyme. More than likely, it was somewhere between 10-20 years ago, and it went undiagnosed for all those years.

I was born with a myopathy (weak muscles), and some minor neurologic issues/deficits. It took me longer to hold up my head, walk, tie my shoes: the normal progressions that all children make. However, I eventually got up to a normal functioning level after years of occupational therapy.

That's where things get fuzzy. Due to the myopathy being a neuromuscular disease, it's nearly impossible to know where the myopathy symptoms end and the Lyme symptoms begin.

All of my life, I have been physically weaker than average, no matter how hard I worked. My minor neurologic deficits would rear their ugly heads on occasion, make certain tasks more difficult, and so on. That was par for the course with me. But, by my early twenties, I started to notice that I was getting weaker, and my symptoms were getting worse. As a licensed veterinary technician, my strength and physical stamina are vital to my job. Becoming weak was a major concern.

By 22, I had been diagnosed with arthritis and ossification throughout my spine, and I underwent a breast reduction surgery to help alleviate some of the back pain I was experiencing. I thought it was rather odd that a twenty two year old would have such back problems since I had not been in any car accidents, but the doctors did not seem to make a fuss over it.

Over the years, I went to doctors,

underwent medical testing of all sorts: blood panels, x-rays, EMG' s, CT scans, ultrasounds, upper endoscopies, colonoscopies, and more. I was diagnosed with gastric ulcers, IBS, and some other minor health issues, but nothing ever explained the progressive weakness.

When I was 21, I could lift a 75 pound dog by myself. By the time I was 26, I struggled to carry 30 pounds. I was also constantly tired. No matter how much I slept, it was never enough. Falling asleep at home, at class, or even driving were a constant issue. Yet, no one had an answer, nor did they have the desire to pursue my case any further. I even had one doctor who was so frustrated by my case, that while on the phone with me, he just hung up – he did not want to be " bothered" by me. I was starting to feel as if I had lost my mind. Could I really be making this up like they said? All my tests always came back normal? What on Earth was going on with me?

Just before my 26th birthday, I was in a minor car accident (the airbags didn' t even deploy. Most people wouldn' t even mention it as an accident, it was so minor). Soon after the accident, I began to have worse and worse back pain. It didn' t take long until I was completely laid up, unable to move due to the pain. My doctor at the time treated me the way most pain patients are treated: heavy duty drugs. I was on Percocet, heavy doses of muscle relaxants, and I literally over-dosed on Tylenol daily (about 3,000 mg per day). I couldn't walk, I couldn't function. I was so drugged, I could hardly do anything. Still, no testing was done to understand why I was having such horrific back pain from such a minor accident.

I started getting regular chiropractic adjustments and having massage therapy done. The chiropractor took more x-rays. The arthritis and ossification was worse – much worse. In fact, my entire neck was just a glob of arthritic bone. There was a large curve in my mid-spine, and it was even detected that my pelvis was rotated forward, so that pulled on my lumbar spine even more. I began an intensive 3 times per week treatment of adjustments and massages. It helped the back pain tremendously, and within a few weeks, I was a functioning human being again. However, the mystery of the cause of my pain and spinal issues still remained a mystery.

In 2008, I began to notice even more pain. Now my hips were constantly aching. My feet hurt easily. I couldn' t write or hold onto any objects for any length of time without my hands causing tremendous pain. By this point, literally every joint in my body hurt, and I could not carry 20 pounds. My General Practitioner sent me to specialists all over. I had more blood, x-rays and EMG' s performed. Everything still came back normal. This doctor, however, refused to give up on me. He told me that he would send me to specialists until we got a diagnosis. Finally, someone who believed me!

After much testing, and many doctors, I was sent to a Rheumatologist. He asked me if I had ever been tested for Lyme. I didn' t recall ever being tested for it, so I told him I didn' t think I had. Now, it' s important to note that I am from Long Island, New York. The East end of Long Island has one of the highest rates of Lyme in the country! I spent time in up-state New York, Pennsylvania, Michigan, Virginia – all high risk areas for Lyme. Not one doctor in any of those states ever thought about testing me for a disease common to the area.

Imagine my surprise when I got the phone call saying that I came back positive for Lyme. Lyme! I never would have expected it. I never found a tick on me. I never had the infamous

" Bull' s Eye Rash." I never had any of the
" classic" signs of Lyme. I immediately called
my mother. We finally had an answer! That
night, I began doing research on Lyme and
treatments. In doing my research I found that
not only was my weakness from the Lyme, but
my gastric ulcers and IBS were also more than
likely caused by it. More than likely, the Lyme
also attacked my joints due to my already
weakened musculature from the myopathy.
Lyme affects different patients differently, no
two cases are alike. It just makes sense that
the bacteria would attack the weakest parts of
my body, thus making them weaker. Suddenly,
everything fit together. All these various pains
and issues caused by one underlying problem:
Borrelia burgdorferi – the Lyme bacteria.

I was placed initially on Doxycycline, an
antibiotic. The problem with the doxy is that
it' s really mainly used to treat newer
infections. Since I had had my Lyme for so
long, the doxy did nothing but make me vomit.
After seeing an infectious disease doctor, I had
a Picc line placed (kind of like a permanent IV
catheter), and I was on IV antibiotics for a
month. Like most patients on the IV antibiotics,
I experienced something called a " Herx
Reaction." A Herxheimer or " Herx" reaction

is when patients actually feel worse as they go through treatment due to the toxins released when the bacteria die. It was a very difficult month for me physically, but I endured the worst.

The Picc line had been out for a few weeks when a new symptom came along: seizures. I began having simple partial motor seizures. Sadly, seizures are not uncommon in Lyme patients. The permanent damage to my bones and joints is irreversible, so I was also battling the fact that I am weaker than a normal 29 year old. I went through countless CT's, MRI's, a spinal tap, tons of blood work and so much more. After months and months of no answers and a worsening condition, I received even worse news: I now had MS.

Though the MS Society does not have an exact pinpoint on a cause of MS, one thing that often comes up is that it is strongly believed that certain viruses and/or bacteria may play a large contributing role in the development of the disease known as MS. One of the bacteria that is considered highly suspect is the Borrelia Burgdorferi: the bacteria that cause Lyme. I cannot officially say that my Lyme has caused my problems with MS, but to call it a coincidence is an understatement. People so often underestimate Lyme, and yet here I am with MS as a result of it. Lyme is a far more potent illness than most people can ever fathom.

I still have constant arthritis pain in various joints and parts of my body. I am still constantly fatigued. I could sleep for 12 hours, and my body still craves more. It can be difficult at times. I could easily be viewed as lazy, but the truth of the matter is: between the exhaustion and arthritis pain, doing even the most basic of tasks can be extremely strenuous for me.

I know that my future holds more Picc lines, more blood work, constant testing and treatments. I also know that my pain will never dissipate, but rather increase. I know that by 35, I could be looking at spinal surgery, knee and/or hip replacements and more. I know that my Lyme is a constant.

There is much debate in the medical world about whether Lyme is chronic, debilitating, can cause MS, or not. I am living proof that it is. There is no way you can look at me, look at my x-rays and blood work; CT' s; MRI' s; spinal tap results, even my symptoms and not see a chronic condition.

Lyme is often down-played as well. Lyme is a debilitating condition, and it needs to be treated as such. Lyme is not a mockery of a diagnosis. It is a real condition that seriously affects its patients.

Despite the pain and chronic nature of my Lyme, the worst news came when I learned that I could no longer donate blood or organs. I've been a blood donor for over 10 years, and am on the organ donor registry in 5 states. I was absolutely crushed when I learned that I could no longer donate – this is a cause that I strongly believe in, and my ability to help others has been taken away from me.

Lyme can be transmitted through blood transfusions. So, my Lyme could be passed onto another person if they received any of my blood. The worst part about that is that I am type O+ the universal donor. Every blood type could receive blood from me, and now that opportunity to give to so many was ruined.

Chronic Lyme can also affect major organs including the kidneys, brain, and even smaller organs like eyes. Since I've had my Lyme for so long, there is no way of knowing what organs and how many have been affected, and how badly they have been affected. For that reason, I cannot risk donating a bad organ, or passing along the Lyme through the organ tissue.

When I was diagnosed with Lyme disease, I had no idea how much it would affect me. I never imagined that it would hinder my ability to donate blood or organs. Lyme attacks the body far worse than most people imagine. When we "Lymies" suffer, others lose out because of our inability to donate much needed blood and organs.

I hope that one day, Lyme is taken seriously. I hope that one day, we will know as much about Lyme as we do so many other conditions. I hope that one day, that reliable testing is available. I hope that one day, that all Lyme patients receive the proper care and treatment. I hope that no one else has to go 20 years without a diagnosis.

Heidi Healy
Bethlehem, Pennsylvania

I ask you what do myself, my husband, my
three children, my mother, my brother, my
brother-in-law, my sister-in-law, my niece and
my good friend from college and her son all
have in common? All of us have or have had
Lyme disease, live or lived in NJ/PA and have
co-infections involved with ticks. People ask
me- How could this be? Logically I know it can
be, because Lyme disease is on the rise, is
very easy to acquire and is misdiagnosed quite
frequently as other diseases.

Please allow me to introduce myself. My name is Heidi Healy and I am a 39 year old female with a long history of strange medical problems dating back to the late nineties. Years before, I had lived in an endemic area for ticks in Flanders, Morris County, NJ. There were various animals that roamed our yard including many deer and mice that all carry ticks. After doing much yard work in the late 90' s in our new home, I had a rash all over my body and my health started to deteriorate. However it was not a bulls eye rash. I was a once healthy active individual with a bright career. I was on the top of my game with a new MBA and a terrific job. I found myself surrounded by strange symptoms. I consulted many doctors in Morris County. My primary care physician thought about Lyme but never pursued it in depth. I trusted my doctor and didn' t question his judgment. He did one test on me and said I was fine. The ELISA test is known to not be very sensitive and misses many Lyme cases.

I went to Infectious Disease doctors who told me I might have Fibromyalgia. I went to gastroenterologists who could only find red spots in my stomach lining. I went to neurologists with sharp pains in my face and tingling. They said it could be stress related. I had pain down my leg and they sent me for physical therapy. I had sharp jaw pain and was given a retainer and some cortisone shots by a TMJ specialist. A surgeon almost performed surgery on me for sharp pains I had in my groin area. A hernia specialist said it was probably tendonitis and not to operate. Nothing ever showed up in CAT scans.

I lived with pain for a while until I met a wonderful doctor in Flemington, NJ. He took one look at all my crazy symptoms on a once healthy active young woman, and thought there was direct clinical evidence for Lyme disease. A urine PCR test came back positive. I was started on treatment for Lyme, but at that point I had probably had it for so long it had gone to the chronic stage. It was deep in my tissues and not floating around in the blood stream.

In my battle to get well, I took numerous oral antibiotics on and off. While some of them may have helped lessen my symptoms, once off them I always relapsed. I had to fight my insurance company to recognize the severity of my disease and have them approve IV therapy. They only approved 28 days of therapy based on an abnormal Brain SPECT scan and a positive PCR of my stomach tissue sample. Can you believe that? After all that oral medication, the disease still thrived and was found by DNA testing in my stomach tissue.

During the first IV therapy some symptoms calmed down and I started to feel better. However, once off the IV therapy, things relapsed again and many new symptoms came out. The encephalitis got worse. I had memory loss, terrible cognitive functions, new facial pains, new knee pains, new athralgias all over and cardiac pains. I ended up having great arthritic and neurological manifestations. This was a case of late/chronic Lyme not an early localized infection. There were new and extending manifestations of the disease. It was even hard for me to drive a car at times because my left leg hurt so much. Since I wasn' t getting any better, I figured it was time to see one of the best known doctors that has researched Lyme and has treated hundreds of Lyme patients. I traveled out to Long Island in April 2005 to meet Dr. Burrascano. I didn' t know what to expect but when I got there an inner peace came over me. Dr. Burrascano looked over my history and spent a lot of time with me. I had never met a doctor who took such care in getting to understand his patient and reviewing all my symptoms. He did more testing on me and I was found to have the Babesia co-infection also. I had probably had it for a while and my antibody level was extremely high at 1280. I remember him commenting- " You must have a great immune system to keep this in place." Babesia is very dangerous and can even lead to death. I am forever grateful to Dr. B for treating me and for his dedication to the Lyme community. Words cannot describe a doctor who truly treats his patient as a human rather than a statistic. Even though I am still struggling with the diseases, I have made great progress and hope to make more in the future. Dr. B has also helped my husband also and for this we will forever be grateful. Without his knowledge and aggressive treatment, I believe my husband and I would be far worse off today.

It' s hard enough to deal with an illness when a mother and father have it but our entire family has this disease. All three of my children have acquired it in different ways and with different co-infections. My oldest son, Ryan got bit twice. The first time my pediatrician told us as long as he didn' t get a bulls eye, he didn' t have Lyme. I know this now to be so untrue. I wish I knew more back then. He got a strange pneumonia shortly after that bite. He had personality changes and mood swings. He had stomach issues and light sensitivities. He had a horrible cough that wouldn't go away at times. My regular pediatricians never suspected anything though I told them I had Lyme. He got bit again and was put on 4 weeks of antibiotics– however new symptoms developed. He became forgetful and got headaches sometimes. I finally called Dr. Jones in Connecticut and he ordered some testing for both my boys. My younger son Timmy showed signs of the disease also, however I never saw him get bit. Tests were done on both and sure enough both boys were CDC positive on the Western Blot test for the Lyme. They believe Timmy may have acquired it from me while I had the active infection and didn't know through the birthing process or breast feeding. Ryan also ended up having the Babesia and Timmy ended up having the Mycoplasma. Dr. Jones took great care with them. He went over everything in their history before deciding on a course of action. Today my kids are doing well. They are good students, they are happy and Timmy scored 10 goals in soccer this past season.!!! I am grateful to have pursued their diagnosis and I am grateful they are doing well. However, I often wonder about the mothers out there that don't know enough about this disease and how their children may not be getting the proper diagnosis and treatment they deserve.

My story doesn' t end there. Since I wasn' t feeling the greatest in 2002 and thought it may be just from going through the process of having kids, I wanted to give my body a break. My husband and I decided we would adopt a third child instead of me going through a pregnancy. We loved kids and I had always dreamed of having a daughter. Shortly before Sarah was to come, I did find out I had the Lyme. I thoroughly expected to recover from the disease and I proceeded with the adoption. At that point, I didn' t realize just how devastating the disease could be. Through my pain and quest to get better, Sarah has been a bright star in my life. She came from China and is such a wonderful gift. It saddens me to say that she even has acquired the disease. During a trip to Cape Cod and Martha' s Vineyard last year she got very sick. She got a high fever, threw up, was lifeless and had a horrible cough. She wouldn' t sleep at night and cried. She said her head hurt and her personality changed. A horrible rash broke out on her chest and back a day after we got home. I immediately took her to our pediatrician who basically stated it was " just viral" and not too worry. However, I knew better. I had had reservations about going up to the Cape because I knew the islands were havens for ticks. I hoped a day trip to Martha' s Vineyard would be ok and we would check the kids extensively for ticks. They had played in a park by the ocean while some of us shopped. The grass was green and beautiful but I' ll always wonder what lurked in it. My child had always been the picture of health and whatever she had just wiped her out. I had her tested through my first Lyme doctor and also through Dr. Jones. She was positive for the Lyme and Bartonella. Once she was put on antibiotics her symptoms changed and she became much healthier.

I have told my story and Lyme will not go away. It is here to stay and we as a community need to stand up and educate the public about the disease. Too many times this disease is trivialized therefore, resulting in a person going on to the chronic stages of the disease. If a person is bit and they get a bulls eye- at least they have recognized the disease and gotten treatment right away. However, what if they don' t know they were bit or knew they were bit and never got the classic bulls eye? There are many cases whereby people just got strange rashes. Many signs and symptoms can be missed therefore propelling the patient into chronic states of the illness. Testing is hit or miss and most doctors don' t take a direct interest in the disease. Insurance companies do not want to pay for long term treatment for the disease. What a mess this disease is! I am amazed that this day and age in America not more is being done to recognize, treat and prevent this disease. It is not going to go away and the population is only going to get sicker.

The new IDS (Infectious Disease Society) guidelines are an insult to the many people who have suffered with this disease. They don' t address the chronic issues associated with the disease and promote meek and mild treatment options. Doctors should be extensively trained to learn about the disease and need to take a more proactive roll in recognizing the signs and symptoms. Doctors should be able to make clinical judgments about patients they believe have the disease since there is no one test that can prove Lyme 100% of the time. The IDS needs to come up with better treatment options because TOO MANY people are suffering. With the new guidelines more patients will fall through the cracks. No one can say Lyme isn' t chronic- look at my case with the stomach tissue sample. I was on orals but my case was so far involved the orals never even treated the spirochetes that were still in my stomach.

Insurance companies need to treat patients better. Recently, all my oral antibiotics have been blocked and my insurance will no longer pay for my meds. They state there is no medical necessity. How can they do this? And to top things off they didn' t even notify me about the block until I found out myself. I' ve had positive tests for the Babesia as recent as July. The Pennsylvania Department of Health has called my home asking questions about my health and how I think I might have gotten the disease. They said they were forwarding my positive tests/statistics to the CDC. I still have many clinical symptoms. The medication to help rid the parasite is called Mepron and costs $800 a bottle. I need 2/3 bottles a month. It is just mind boggling to me that the insurance company can turn around and just block my meds. I even had a doctor within the insurance company' s network agree with me that I still needed treatment along with 3 other doctors. They totally disregarded all my physicians letters and what is best for my health and will not pay anymore. This is an injustice and it needs to be addressed. I feel the new IDSA guidelines have created more of an environment where insurance can get away without paying for treatments and doctors can get away with not addressing the disease fully. There is no short answer to Lyme. The new guidelines treat the disease as if it' s not persistent. They do not look at the true science behind the disease. These guidelines are not science based because if the doctors who truly wrote these guidelines understood the complexity of the disease they would not be stating you have to have a bulls eye and 2 weeks of treatment is enough. These guidelines need to be changed and our nation has to have better care for Lyme patients. The problem is only going to get worse and more adults and especially children will suffer needlessly. I urge anyone who lives in an epidemic area and has strange symptoms to find a doctor that is very knowledgeable in Lyme, read the ILADS (International Lyme and Associated Diseases Society) Guidelines, go to www.lymenet.org and learn as much as possible. Thanks for your time.

Shandy Monte
Fall River, Massachusetts

It is a hard living in a body you don't recognize.
I look in the mirror and I never see myself. I
see someone I do not know. I am trapped in my
own body, and I have no control over it. I miss
my former self, I miss my life, I miss the person
I used to be. I was exciting, ambitious, talented,
loving, energetic and FULL of life.... I am a
shell of myself. I have lost a lot and I guess I
have gained a lot (including weight) lol. People
in my life who I thought were my very good
friends walked away from me⋯ Even family. I
am always scared the end is near and I will

never get to marry the love of my life, or be here to be a mom to my amazing children.

It all started in 2006. Life couldn't be any better at that point. I was doing great in college, in a fabulous relationship and planning my future. Everyday was a blessing. I had never been happier in my life than I was at this time. I woke up with a smile on my face and with a feeling in my heart that FINALLY all my dreams were coming true. I had a great group of friends. I had it all... or so I thought.

Joey and I met 7 years ago while doing Jesus Christ Superstar. I fell in love instantly... and our love of performing brought us together for what has become the most important relationship of my life. One night, my fiancé and I while doing a mini performance of a show we were part of. I remember my throat not feeling quite right. After the performance, I started getting itchy all over. By the next day, I had a blazing red rash all over my body. My thighs were covered in a sheet of red, my arms, buttocks, and stomach. It was so itchy and burned so badly the shower was torture.

I went to 4 different doctors who all said it was nothing. I believed them and went on my way.

The rash went away after a few weeks. I always wondered where and how I got that rash.

In June things started getting very weird for me. My life changed over night and I did not know why. I woke up with overwhelming dread and despair. I lost myself. My life was falling apart right before my very eyes. I couldn't breath, my heart would race and bed became my place of safety. I dropped out of school, tried to leave my boyfriend and could no longer be involved with friends or my son's activities. I would look in the mirror and see a stranger looking back at me. I was a hermit in a dark shell. I couldn't find my way out. I was brought to a local hospital and diagnosed with panic disorder and given Celexa. I cried everyday for months. I never felt the same from that day on.

After working out with a trainer at the gym one day, I thought I would die for sure. I could not make it up the stairs to get to my car. I could not make it into my house. My body was as heavy as lead. I called my fiancé frantic... I literally thought I was dying. This is not a normal response to exercise, I thought to myself. I have been working out 5 days a week for years. Why this sudden extreme response??? I lay in bed for 2 days straight in pain and with what felt like a boulder had been placed over my body. What the hell was happening to me?

In August of 2006, I found out I was pregnant. I was happy but very scared at the same time. I knew something was not right with me, and now I had a baby growing inside of the body that was no longer something that was familiar to me. My fiancé was so happy, and with his happiness came my feeling of safety. I would get through this, hey maybe it will make me better... Was I wrong···? SO wrong.

I spent the next 3 months in the house and only told a certain few of my pregnancy. The first 3 months of my pregnancy were unlike my pregnancy with my son. This just didn't feel right, but I carried on. I had weird dizziness, and my morning sickness was more like day sickness. When I would go places it was as if I was looking through a screen... I wasn't really there. My doctors assured me it was all hormones. I arrived at my second trimester and although I still had morning sickness I was feeling better. I was growing rapidly and gained a lot of weight very quickly. It was hard on my small frame. When I was 5 months I look more like I was 9 months!

I started having complications with early
contractions and than later found out I had what
is called POLYHYDRAMNIOS. Polyhydramnios
is when there is too much amniotic fluid around
the baby. They do not know what causes this
(what else is new) and I had to now go and
have weekly ultrasounds to make sure
everything was ok with the baby. Talk about
added stress on and already stressed body.
They said that having this condition could mean
that my baby's kidneys were not functioning
properly or that she had a cleft palpate... I
about lost it. I couldn't deal with what was
happening to me, and now there could be
something wrong with my baby. I was on bed
rest from month 6 on and I could barely walk.

My third trimester is when all hell broke loose
in my body. I had so many symptoms unrelated
to pregnancy it was quite ridiculous··· Yet no
one would listen. I had stabbing headaches on
the right side of my head. It felt like someone
was stabbing me with a knife. They were very
scary. I had dizziness constantly and to my
surprise couldn't walk into stores anymore. My
heart rate was in the 150's constantly and I had
to be monitored with event and holter heart

monitors. I was also worked up at the hospital many times. All to find "Nothing wrong"... It's just hormones... Why did I feel like my life was ending... why did I feel like something was seriously WRONG??? Even my fiancé started to think I was being a hypochondriac. He is my greatest support but when you have top notch doctors telling you over and over nothing is wrong. The favor goes to the other side.

I lay in bed everyday during my 8th and 9th month. I would leave bed to eat and go to the bathroom. Crying everyday if I would make it through this to see my baby, to marry my fiancé, to watch my son grow up. Than March 21st 2007.... Our lives changed...

Lying in bed, my water broke and woke me from my already broken sleep due to a racing heart. I thought this is it! We frantically called my doc because with Polyhydramnios, the fluid comes out so quickly, because there is so much. There is a risk of the baby's cord falling out... and this is NOT good. We got to the hospital quickly. I was set up and the contractions were instant. It was a 5 hour labor and wiped me completely out. I remember when my daughter was getting cleaned and everyone was taking pictures and congratulating. I felt like I was not there in the room. I felt separated from everyone else... like a wall was between us.

This is where my life changed for the better and worse. My beautiful daughter was here at last; however, my ability to be her mom had been ripped away from me before Joey could cut her umbilical cord. At that time the most joyous moment of my life, became the most fearful.

My daughter had some breathing difficulties and had to be put into level 2 nursery··· another stress. I tried to breast feed, but I was so weak. I had to use a wheel chair in the hospital because of weakness and dizziness. I couldn't sleep. I had anxiety, heart palpitations and head pressure. By the last day in the hospital I passed out in the nursery. They sent me home without my baby. They sent me home scared with no answers as to what was happening to me.

Things continued to get worse and worse. I had so many symptoms I could not count them all. My baby wasn't with us, and my fiancé was confused as to what was happening to me. I called my doctor from home and told him···

" My lips are numb, my head is tingling, I am dizzy and I cannot walk... I cannot feel my fingers, WHAT is WRONG with me???" He had no answer but to go to the ER.

Days turned into weeks, weeks turned into months and months turned into a year and STILL I was sick. Not a day went by that I felt well... things actually got worse. Fourteen months of my life and the first year of my daughter' s life would be spent in ER's, Doctors offices 4x a week, specialists, psychics, card readers, and medical intuitives. Desperation had set it, we needed answers.

I missed out on my daughters first year of life, so sick I thought I would surely not make it. I was diagnosed with a plethora of diseases. Lupus, MS, Sarcoidosis, and of course I was a mental case and there was NOTHING wrong... Yet I could not function. I became bedridden and needed assistance with everything. My fiancé lost his job because he had to care for me. We were getting no answers. We were scared and helpless... and there were kids involved. I cried everyday because I wanted to be like other moms taking their babies for walks. I would hear the moms walking by my house and I would just break down and cry...

Why is this happening to me, what did I do wrong. I would say over and over "all I did was have a baby".

I started researching and basically diagnosed myself with Lyme. I had every symptom of Lyme and than some. I found a LLMD (Lyme literate medical doctor) who tested me for Lyme thorough IgeneX. He found me to have 5 positive bands on the Gig western blot and also a low CD57, which is believed to only be lowered by the Lyme bacterium.

I started on antibiotics in May 2008. I was finally diagnosed with Neuroborreliosis (neurological Lyme disease) 14 months after my daughter was born. After being bedridden and unable to do anything for myself, my children and my fiancé. I may have had this for years dormant in my body and never knew it. I filmed movies in the woods, went hiking, went camping every year as a child and even had a tick attachment at the age of 12. The trauma from my pregnancy is thought to have brought this out full force.

I am no where close to functional yet, but I am not completely bedridden anymore. I am homebound and couch bound. I go out for short car rides and sometimes just to sit on the beach. I still suffer with a tremendous amount of symptoms. And I still cannot be the active mom I want to be.

I have the most amazing man by my side that has taken care of me every step of the way and NEVER complained. He is my rock, my safety, my friend, my lover and my strength. My children, my family are my reasons to live and to fight, no matter how many times I have just wanted to give up.

One thing Joey and I miss dearly is performing. It was a huge part of our lives and part of our spirit. My love of theatre and film, performing, singing, acting... It is a part of me that I have lost. I also hold onto the hope that I will be back to doing what I love again.

To top it all off, my son was also diagnosed with Lyme disease. He was very sick but thank god is doing great on antibiotics. He is back in school and back to his activities. As I write this, to my surprise, my little brother was also brought into the doctors a few days back and the immediate response was "Lyme disease". This seems to be a family trait, one we can surely do without.

This has my family thinking. Could this have been congenitally passed on by my mom, who for many years (Over a decade) has struggled and suffered with diagnoses such as "Fibromyalgia", "CFS", "arthritis", "Depression", "Anxiety", and a plethora of many other symptoms... along with "there is nothing wrong with you, it is all in your head". This seems to be the trend when it comes to Lyme disease. All this time I have wondered HOW I got this. Maybe the answer has been in front of me the whole time... my Mom. My fiancé also shows signs of Lyme disease and also has many exposures. It just shows you how widespread and epidemic this illness really is. I would never have believed this, until it happened to me and my family.

My only hope is to get well, so I can be a mom again, give Rayn (my daughter) a chance to know the mom that my Mikey (my son) knows and misses so much. To be the woman my fiancé fell in love with. To be ME again.

Tests:
IgG Western blot Positive 5 bands and 2 IND, IgM negative with one IND band
CD57= 43 (range 60-200)

White Blood Counts are elevated
CRP is high (inflammation)
Rheumatoid factor high (has gone down a little
with antibiotics)
ANA + (negative when on antibiotics)

Symptoms:
Severe Head Pressure
constant dizziness for 18 months straight
veering to the left
vertigo
the feeling of being on a boat for 18 months
straight
heart palpitations
heart racing
muscle pain
knee pain
feeling drunk or toxic
foot pain
tingling feelings
fatigue
insomnia
FEAR
Anxiety
floaters and blurry vision
streaks on my body
agitation
anger
crying fits

depression
deconditioning from being bedridden so long
and more....

It truly was unexpected.

Carla Ashway
Newburg, PA

There is a time in your life when you contract Lyme disease that your normal life ends and a new life begins. This is my Lyme story and how it has changed my life.

Before I contracted Lyme disease I was a very healthy 29-year-old. I grew up in Franklin County PA in a very rural area. I loved spending time outdoors, spending time with friends and family, and I was always very active. I worked as a veterinary technician for a few years before moving to CA in 2001. My father had passed away in 1998 due to cancer and I just needed to get away for a while and wanted to start a new life. In CA I worked as a fashion fitting model for a fashion designer and I also had my own business.

In 2006, I was missing home so I moved back to my childhood home in PA.

In early April of 2008, my life changed. Everything went from normal day to day life to misery in the blink of an eye. One afternoon I started feeling a weird sensation in my back as if someone had applied a muscle cream to my back. It was one of the strangest sensations I have ever felt. It was an icy-hot sensation that would not go away. I thought I had pulled a muscle or something and decided to lye down for a while because I was starting to not feel very well. While I was lying in bed, I started to feel these icy-hot sensations moving down my arms and legs. I got really scared and wondered if I had been bitten by a spider. It was almost as if I could feel a poison moving throughout my body. It then started to move up the back of my head and across my scalp. Later that evening I started to run a fever, had swollen lymph nodes in the back of head and neck, and also developed severe GI upset. I was also trembling and could tell that I had a rapid heartbeat. My appetite was gone and I could not eat. I had severe fatigue as well. I figured I had the flu or something so I decided to go to the doctor in the morning.

My primary physician was closed for the day so I went to a walk in family practice. The doctor examined me and listened to my symptoms. He wrote it off as an upper respiratory infection and gave me 7 days of Avelox.

At home I continued to feel worse and worse and I was not getting any better. I was sleeping close to 20 hours a day. I was also had a lot of anxiety for the first time in my life. I decided to go to the ER later that week because my heart was racing and I wanted to make sure I was okay. Nothing had ever happened to me like this before.

At the emergency room I was lucky enough to see a doctor that used to be my family doctor when I was younger. He ran blood work, a urinalysis, chest x-ray, and also an EKG. I was also hooked up to a heart monitor. While sitting in the bed, my heart rate was 130. I also asked for a Lyme titer and a mono test. Everything came back normal except for a slight elevation in my white blood cell count. The Lyme titer was also negative. While sitting in the ER, my mother and fiancé noticed a small red bump on the back of my head at my hairline. The doctor looked at it as well and but thought I had a virus of some sort. He told me to continue my Avelox and if I did not get well, to see my primary doctor.

I few days later I started to develop blurry vision and tingling sensations in my scalp. I went to see my primary doctor and he thought maybe I had some sort of Strep throat or something that the Avelox was not helping. I asked him about the GI upset and loss of appetite. He did not have any answers for me. He decided to give me 500 mg of amoxicillin 3 times a day for 10 days. I thought this was the answer to my problems because amoxicillin always helped me as a child when I was sick. At last I would feel better.

A day after starting the amoxicillin I got even sicker. It felt like fire running through my veins. I was very hot and boiling up. My upper eyelids were swollen and my GI tract was even worse. My heart continued to race and pound. My vision was blurry and I could not eat. It would take me close to an hour just to get down a banana or some Jello. I was also sleeping all the time and was very scared to get out of bed.

By this time it was around the end of April and I was still sick. The amoxicillin was not helping and it was making me worse. Again I went back to the ER. This time I saw a different doctor who really did not want to be bothered with me because I guess I did not look sick to him. I was still running a temperature, had no appetite, and had swollen glands, blurry vision, and tachycardia. The doctor ran blood work and also 2 blood cultures in addition to another chest x-ray. He stormed into the room after the results were in and basically told me the ER was not a place for someone like me to be since it was not an emergency situation. He also told me I should be taking my problems to my family doctor and not the ER. Again, I was told it was a virus and sent on my way.
By this time it was May. Spring was here and I was miserable. I was in bed all the time and feeling terrible. My symptoms were not going away and the anxiety of not knowing what was wrong with me was really starting to get to me.

I continued to see my family doctor several more times and he could never give me a straight answer as to what was wrong with me.

I demanded more tests be done because I was so tired of being sick. I was sent for more blood work including a West Nile virus test, another mono test, and also I was checked for C. diff and parasites due to the severe GI upset I was having. Everything came back negative.

I was starting to think I was never going to feel well again. The left side of my face started to tingle and I still had tingling sensations all through my scalp. My symptoms continued and I still could not eat. I decided that if I did not feel any better by the next day, I was going to go to a different hospital in the hopes of getting a diagnosis and an answer as to what was wrong with me. My fiancé drove me to a different hospital and I was seen again.

I had been doing research online looking for an answer to my problems. I knew I had either mono or Lyme disease. I mentioned these to the doctor and he said, " You do not display the symptoms of Lyme disease." I said, " What about mono?" Again he told me that I did not have mono and I did not display the characteristics of mono. He did more blood work and told me everything came back normal. I asked him what was causing the tingling in my face and head and he said, " Paresthesias in the face are always caused by hyperventilation." I told him that I have never hyperventilated in my life. He just shrugged his shoulders and gave me a blank look because he had no idea what was wrong with me. After he left the room I broke down in tears in front of my Mom and fiancé. I was getting so sick of spending thousands of dollars and not getting any answers. I asked for a copy of my labs that the doctor said was normal. It turns out they were not normal. I had an elevated white blood cell count and also my lymphocytes were slightly elevated.

My mom saw how much distress I was in and that I was on the verge of breaking down because I could not get any help or a diagnosis. She called across the street to her family doctor who agreed to see me.

Again, I explained all of my symptoms and he listened to what I had to say. He and I agreed that I should have a CAT scan of my brain and an ultrasound of my abdomen to see if there was anything wrong that could be causing all of these problems. The CAT scan of my brain came back normal and the ultrasound of my abdomen was normal as well. They did find that I had only kidney but this was no surprise to me

since my mother and grandmother were also only born with one.
This was now the end of May and I was still sick and depressed. I still did not have any answers. The new family doctor decided to send me to a cardiologist next. I sat around for another month until I could get in to see the cardiologist in June.

It was now a few days before my 30th birthday and I would have good days and bad days. Some days I could function and did not have my symptoms. Other days I was very tired and did not feel well.

The cardiologist gave me a 24 hour halter monitor and also an echo of my heart. The echo was normal except for some trace MVP. The halter monitor was also normal. The cardiologist was not concerned at all about the tachycardia. He said I was young and it would not hurt me. He thought maybe I had an Epstein-Barr type virus or something.

Now it was July. I has spent most of my 30th birthday in bed crying and depressed because I was scared out of my mind and I did not know what was wrong with me. I continued to see my new family doctor and was also sent to an Ear,

Nose, and Throat doctor. The Ear, Nose, and Throat doctor had no clue what was wrong with me. I had a white tongue for months that I suspected was thrush. I showed him my tongue and he said, " I see so many tongues and your tongue is nothing to worry about." He gave me some nasal spray and sent me on my way.

Again I went back to my family doctor. I showed him my white tongue and he simply shrugged his shoulders and said, " So what." That day I also took in a list of Lyme disease symptoms for him to look at. I had highlighted all of my symptoms on the list and I had just about everything on the list. I was also having muscle fasciculations and stabbing muscle pains by this point. While sitting in his office, he could see the muscles in my legs jumping around. I was also messing up my words when I spoke and my heart rate was still on the high side. We talked about Lyme disease and he argued with me about it telling me that I did not have Lyme disease. He agreed to give me yet another Lyme titer so we could put a close on the issue. He said, ' If you take enough of these Lyme titers, it is like flipping a coin. Sooner or later you are going to get a positive even though you do not have it." He insisted I did not have Lyme disease and that I need to stay off the internet and stop diagnosing myself.

I got a call a few days later that the Lyme titer was negative. I broke down at this point. What was wrong with me? If I did not have Lyme, then what was it? Anxiety was overtaking me. I was afraid to be left alone and I thought I was going to die unless someone helped me soon. I have always been an independent person and for me to act this way was not in my nature.

I asked to see a neurologist in the hopes of finding out what was causing the tingling in my head and face. The neurologist asked me if I had any joint pain. I said no and he dismissed Lyme disease. He gave me a full neurological exam and said he felt I did not have MS or anything serious. He thought maybe I had some sort of nerve damage or something. He gave me a script for some sort of antidepressant that was to heal nerve damage. When I walked out of his office I knew that I was not going to take the medicine because this was not the answer to my problems. It would only mask the real problem.

I was at my breaking point by the end of July. I would lie on the sofa all day and my muscles would twitch and jump constantly. I was still having off and on fevers and tachycardia. I was also having bouts of anorexia and could not eat. The anxiety was terrible. What was wrong with me?

My fiancé saw how much I was mentally upset by all of this and we researched and researched online. The best possible match to my symptoms was Lyme disease. We went online and found a Lyme literate doctor near Baltimore. He called and got an appointment for me. At last I thought I would know for sure if this was Lyme and hopefully get some help.

I saw my Lyme doctor at the end of July for a consultation. I was so happy. She spent over 2 hours with me just listening to my symptoms and talking to me about everything that had happened. I have never had a doctor spend this much time with me yet alone listen to me. By the end of my appointment she said, " I am 99.75% sure you have Lyme disease." I had blood work drawn that was sent to Igenex for testing.

At this point I was somewhat relieved that I was getting somewhere and I would be feeling better soon. I was still nervous to find out my results. What if my doctor was wrong? What if it was not Lyme? I worried day and night until I got my results at my LLMD's office on August 18, 2008.

I tested IgM positive on the Igenex Western Blot. I also had a few positive bands on the IgG Western Blot. My Babesia Microti test was positive and I was also positive for the Babesia FISH test.

Finally. Finally I had an answer. My doctor began treating me for systemic yeast that has spread through my body due to the negligence of the previous doctors I had seen. I could not immediately begin treatment because the yeast was so bad and taking antibiotics would only make it worse. She also started me on vitamins and supplements to build up my immune system.

I continued to see my LLMD for the next few months working on treating my yeast and also detoxing my body before treatment.

My fiancé started to show symptoms of Lyme disease and tested positive for Lyme and Babesia Duncani.

In October we were finally ready to start treatment for Babesia. I started taking

Azithromycin and Mepron in October but had to stop due to GI upset and then eventually pain around my heart and tachycardia.

I am now on treatment again for Babesia and taking it one day at a time.

Most days I feel terrible and I am unable to do much of anything. My life has been completely turned upside down. I can no longer do any of the things I used to do. My family does not understand what I am going through and cannot understand why I am not better by now.

I have spent close to $50,000 in medical bills that were not covered by insurance. All of this could have been avoided if I was properly diagnosed in April. All of the doctors I saw through my long road to a diagnosis knew nothing about Lyme disease and refused to treat me since I did not have a bulls eye or positive test. They also did not know what coinfections were. If I mention to a doctor that I have Lyme disease and Babesia they say, " Oh, where did you go to get that?" They act as if I went to Mars and came back with this. Pennsylvania is ranked #2 in the nation for the most cases of Lyme disease. There is no excuse for their lack of knowledge and ignorance on Lyme disease. I feel Lyme patients are being treated very inhumanly mainly due to the IDSA and lack of knowledge shown my doctors when I a patient presents with Lyme symptoms.

I still do not know when or where I contracted Lyme and Babesia. I am suspecting I got it at my house. I live in a very rural, wooded area of PA. Ticks are everywhere in my yard. My best guess is I was outside trimming large pine trees in my yard a few days before I became symptomatic. It is funny how something as simple as doing yard work can make your life into something like this.

At this point in my life I am struggling to make it from one day to the next. I no longer remember what being normal is like. I cannot do any of the things I used to do. My life has been completely turned upside down and I want my life back. I want it back for many anything. I will get it back. I will not let this disease have my life.

I look to my father for inspiration to continue fighting. My father fought cancer for 8 very long years before he passed away in 1998. He went through numerous surgeries including the removal of his right lung. He endured years and years of pain, chemotherapy, more surgeries, more chemotherapy, and sickness that was hard to even watch yet endure. He never once complained or showed any sign of giving up. He was the strongest person I have ever known. If my father could fight and not give up and remain strong, I can too.

I will be strong and I will get my life back.

Linda
Fall River, Massachusetts

I am a mother of two beautiful girls. They are wonderful smart and talented. My oldest is 13 and my little one 4. They are great kids, even though at times challenging. I like to going online and doing research when I have the time. My little one keeps me on my toes.

I am half Italian, but grew up with Italian family, and traditions. I am also French Canadian, and Jewish. Of course, I am American. I have always been easy going. But I do have a bit of an Italian temper. My kids are my life now because I really don't have energy for too much else. I need to cook, clean, and that doesn't leave too much energy for fun. I try to make things fun.

I have had about every symptom you can think that goes with Lyme. I always feel better in the summer. But in the winter my symptoms come back worst, and I have more. When it rains, especially when it is cold outside I feel like an 80 year old cripple woman.

When you have Lyme disease as long as I have and you go undiagnosed as I was, you learn to force yourself through your bad days. Cannot get through your worst days, and your good days are bad compared with someone who doesn't have Lyme Disease. So you enjoy your good days.

Sometimes people die from Lyme disease, because once it becomes chronic it is hard to cure, and gets into all your organs including vital. Some people die from heart attacks, others from kidney or liver failure, and lung failure. Others live with it for the rest of there lives. Because of lack of treatment. Anyone can get long term antibiotics for acne thought. Just not for Lyme Disease. Doctors also don't prescribe high enough doses of antibiotic for long term Lyme disease to be cured.

The longer you go without diagnosis the longer it will take for antibiotic treatment to heal you. As most doctors won't treat it you can end up with other auto-immune disease which can kill you too. Most people don't know the seriousness of this illness. You can have it and never feel the bite and never get the bull's eyes rash and doctors won't test you for it. In addition to this; test are inaccurate so you will be told you don't have it. It may take months or years to finally get diagnosed at this point you need IV antibiotics to get cured and or long term oral antibiotics. You can end up with permanent damage. Doctors do not treat it correctly and are not educated about this disease. I am one of many misdiagnosed because of incompetent doctors. I did finally find a good nurse who ran test and found out I had it.

I was lucky my Elise test and Western Blot test were positive. Many are not so lucky. But before I got tested, I was refused testing, told by several doctors that there was nothing wrong with me, and they where not going to waste my insurance company's money to test for anything. I was subsequently, left to deteriorate with multiple illnesses. I also got sick to the point were my Liver was being attacked by the Lyme and by my own immune system.

I just got more and more and more symptoms, and I guess my doctors thought I was a hypochondriac or something. I thought I was going crazy too. Because I just felt so much pain yet doctors told me I was just depressed. I was like if it is depression why can't I walk any more without severe pain. I was walking three miles at least three days a week before.
Then at some point after seeing rheumatologist specialist north of Boston, I was told it was either Fibromyalgia or a manifestation of my imagination. The Rheumatologist in Fall River said there was no way I have Lyme disease or my knees would swell up. Well they didn't.
Then I was misdiagnosed with Fibromyalgia.

I was in tears walking my daughter to school, and it was only four or five blocks away. In the winter the cold would rip through my bones, and the pain was excruciating. One year was particularly difficult for me before I was diagnosed, I had trouble moving. My boyfriend had to lend me his car and I still could not get my daughter to school on time all the time. She missed like 19 days of school that year and I was getting letters that she couldn't miss anymore days. He ended up taking her to school for me and I was able to pick her up in the afternoon after sleeping the whole day away.

I was a human guinea pig, doctors put me on every anti-depressant you could think of but it didn't help. I was just getting worse, because the Lyme disease was multiplying inside my entire body throughout all my organs including vital. I did have anxiety but who wouldn't and part of it was medications.

I saw a great Pulmonary specialist here in Fall River; I think it was my fifth year with Lyme and no treatment except two months of antibiotics for severe sinus infection after a year of neurological Lyme. Anyway had trouble breathing but my test for lungs came back fine. She said it looks like asthma but not bad or showing because it was only after I would be active like doing a load of laundry and going up and down three flights of stairs. But I think she knew I tested positive for lyme. She didn't run any test for co-infections but put me on an anti-malaria medication Quinine I think. I took it for two weeks and my symptoms disappeared. They come and go. They just came back a few months ago but not as bad.

Doctor now said it is bronchitis and asthma and I am taking medication for those now. Ok Great, my doctors prescribed me with a steroid based inhaler which is bad for my compromised Immune system and Lyme. My entire body would be shaking, when I was taking over the counter allergy medication while on different meds. I figured it out when I had same symptoms and went to hospital last year for dehydration··· That is what it was dehydration. That goes with lyme too, and is caused by the kind of allergy medication I took. I felt like I was going to pass out. I was told it was anxiety at emergency room other times I went.

I have chemical sensitivity too. I tried going on pain meds and got so sick. The medication was an anti-seizure medication and they use it for pain there is no way I can take it. I could not move or get out of bed at all on that stuff. It also felt like poison was going through my body. Unless it was from my liver. But I had it again when I tried other medications.

I can't remember all the symptoms they still

come back sometimes. Like this morning and afternoon. I am so exhausted and have a severe headache. It just hurts in my entire head. But not bad like before I got on long term antibiotics. I had all these symptoms every day before without any let up.

I can't stay out in the sun without feeling dehydrated, dizzy and getting migraines for longer than an hour or two or I also feel like I will pass out. Before I could not go in the sun at all. Furthermore, I got low vitamin D levels. Last year they were the second lowest my doctor had ever seen. I read it is one of the symptoms LLMD use to diagnose lyme along with 57cd test which I have never had as far as I know.

I don't get burning pain anymore with headaches, but even my face hurts, and loud noises just go through my head and I am dizzy. I can't stand the light still when I have a headache and my vision gets all blurry sometimes I get aura with them. My hearing is not as good as it was, and I am always saying, what. I need reading glasses now. My head is sore everywhere. I had, have TMJ comes and goes. As well carpel tunnel syndrome, vertigo, stabbing pains, Sharp pains going up my head

and down my neck, back, and shoulder and arm.
I can not turn my head when this happens or
move for a week. All these things come and go
and I don't know what to expect. But it is not
every day and it is not as bad.

It is not all the symptoms at once like before. I
get different ones. They don't last as long
anymore either. I have days where I don't have
any symptoms except the fatigue, which causes
me not to think clearly. It is very difficult to
concentrate and read quite often. I still cannot
remember things sometimes. I was diagnosed
with Restless Leg Syndrome. I had irritable
bowel syndrome but I only get that like once a
month for a day or so now.

I have been I have bursitis, and I get it in my
hip and side. So bad I cannot move without the
sharp pains and stiff and sore for a week. I had
rib pain so bad it felt like the pain you get when
your ribs separate with your first pregnancy.

I can not sit too long or stand to long to this
day. If I walk too much I am in to much pain
afterwards. My oldest used to sit on my lap,
and I had to tell her not to because it hurt so
badly when she was just five. I now still feel
that way sometimes with my little one. I feel so

bad, and mean because I have to say I can't cuddle this way.

I acquired the neurological symptoms first, and then I got allergies, and every joint hurt. The fourth year into my illness I could no longer carry a grocery bag with just milk and soda in it. It felt like my joints were breaking and coming apart the pain was so intense.

My fingers swelled up in the beginning, and I could get my ring off until I kept my hand in a bowl of Ice for an hour and put miracle whip on my finger and it didn't come off easy or without pain.

I have weird shaking in body sometimes. When I sleep and my entire body will jump and wake me up. My body isn't shaking outside but I feel it inside. I was told I have seizure like activity in my brain, but until I pass out from it they cannot diagnose me with actual seizures. I was told I don't have sleep apnea, yet I sometimes wake up gasping for air.

People don't realize when I say I am tired sometimes, I am not just tired. I am weak. It is like you just ran a marathon or walked at least three miles two days in row when you are not used to it.

When you get tired like this it is very difficult, I can't go to sleep because I have to take of my little one. I think I will be taking her to the park in a little while if it is not too hot. I can't go out in the heat or I could have asthma attack or just feel like passing out. I could in no way get out this morning. I had to put my youngest daughter in preschool when she was three. Just so she could get out like a normal child early in the day.

I was so tired I forgot what I was writing above this is another symptom of lyme. I was going to write, that last week I had to wash my hallway and steps because no one else in the building can, they work. I could not see letting it go another day. So I was not using my brain due to the severe fatigue brought on from the week before.

I accidentally put the soap on the hallway floor and stairs and walked across it to put soap on the downstairs hallway floor. Low and behold I slipped and flew down the stairs banging my head so hard my daughter thought the AC fell out the window. It was the back of my head hitting the stairs. lol Ouch. I am so lucky; I only had a sore head and a couple of tiny bruises.

This is life with Lyme Disease after it is not diagnosed for years or treated. Then not treated properly. I think my first treatment was only ten days not two weeks with what was called strong antibiotics time released or something. BUT NO IV. I have read over and over again it should be treated for at least three months once it gets into the brain as mine did. Also, I have not had any stress test on my heart which I believe should be done with Lyme. I did have my heart ultrasound like five years ago and it looked ok or so I was told.

It is very hard to trust doctors after what I have been through.
It just really wiped me out writing this, and my head hurts too much today. I am on antibiotics, same dose as it was for Lyme for the acne I believe the lyme has given me since I never had any before. So I think I am safe for now. Even though it is oral and I can't take a full dose every day.

I tested Positive for Epstein Barr Virus too. My cognitive function is very bad at times. I can't even spell words, and I was always great at that in school. I had to go back and fix a lot of mistakes above but I don't know if I got them all. Above I misspelled soap as soup and didn't notice it, but then I did and then fixed it. It just takes me longer to get things correct. I also was missing words, and had to add them. My writing didn't make sense.

I can't comprehend what I have just read. I don't retain any of the information in my brain. Sometimes I can, but I don't know which days, hours, or minutes. Or I just can't focus to read. Even simple math problems like times tables, I do know are not in my head, when I need the answer. I even forget what day it is sometimes. I forget the actual date.

Thinking back upon my childhood and early as young adult I think I had Lyme disease then. I read you can have Lyme disease and only have a few symptoms. I remember I had a few tick on me growing up, but never knew what they were. I had learning disabilities as a child. I had to have extra help for reading and I would just space out at times.

We used to go to many parks, pond, and beaches with woods nearby. We went to Cape Cod, a known epidemic area for ticks that carry lyme. My family was unaware of Lyme disease as many people and doctors still are clueless. I also believed I had ADHD symptoms as a child and young adult I had so much energy and trouble sitting still to do school work.

I am one of thousands. 10 years I still have symptoms. Due to ignorance of our MD's and improper treatment in the beginning the Lyme disease becomes stronger and harder to kill. Without a change many more will suffer. Without awareness and knowledge more people will become disabled and some will even die. The MD's that do understand the magnitude of this disease have there hands tied do to politics of this disease. Treatment cost keep many from ever getting proper treatment, Insurance companies don't have to cover treatment needed to cure Lyme, and unlike other disease it is improperly treated and undiagnosed. If it were any other disease, patients would have choices of treatment and would have the opportunity to get well but not with Lyme Disease. Lyme disease patients are the only ones told that you' re not sick. It is all in your head. Go to a psychiatrist, Go home your not that bad. Many die after being told that due to organ failure from organs being attacked being tuned away. Had I stayed with listened to doctors and not sought help elsewhere. I would be dead today.

Lyme disease attacked my lungs and liver. Emergency rooms sent me home and did nothing. It took too many years to get diagnosed and I may never get better. Like thousands I may get worst or even die unless I can get to LLMD and get treatment that is correct. Still it could be too late. This is reality for too many. It can happen to anyone, anytime. Be aware, educate yourself protect yourself and family. Because we are dismissed by our doctors our friends and family are not even supportive. They still don't believe it is that bad or doctors would put you in the hospital. With lyme even if you are about to die they just send you home. This is the truth. This is the reality. NO ONE IS SAFE. Lyme is on the rise at a rate higher than that of Aids, West Nile Virus, and Avian Flu combined. Open your Eyes- don' t be clueless.

Jason Hulko
Tamaqua, Pennsylvania

My story begins in 2004. I was a private in the U.S. Army. I graduated 5th highest in my medical class of 386 soldiers. I was honored to work with the top doctors on the Critical Care Burn Ward floor of Brooke Army Medical Center in San Antonio, TX which is the top critical care center for burn patients in the world. I was already at the level of assisting the doctors with skin grafts. I was assured that I would be fast tracked onto the " Green to Gold" program which would have taken me from being an enlisted medic to giving me the chance to become an officer as either a Doctor or a Nurse.

Prior to that, in the last week of our basic training, we did Field Training Exercise, or F.T.X. This involved carrying a 35 lb. pack on a 7 mile hike, sleeping in what they called " hasty trenches" , which is digging a hole in the ground to sleep in, and 10-12 hours of training in a densely wooded area in Georgia for 5 days. It was a few days after our return that I noticed a small rash on my right leg, and a much larger one on the left side of my chest. I originally chalked it up to the fact that I had only been able to change my clothes once in the five days out there or possibly I had stumbled into something I was allergic to. It did seem strange to me that both of them were a splotch in the center surrounded by a ring. Another two soldiers in my platoon had a similar rash and when we questioned our Troop Medical Clinic, they told us it was nothing and to go back to training. We also discussed the fact that we had been given routine weekly injections over the course of a few weeks and when we asked what they were injecting us with, we were told they were just vitamins and we were then punished with long sets of sit ups and pushups for questioning their orders.

I now know that they were classic bull' s eye rashes. Right after, I had one of the worst flu' s that I' ve ever had. It did not come with nausea, but the aches and pains were unbearable, and they have never truly gone away. I now know that this was my first exposure to Lyme. I am not sure if it was from the sleeping in the woods in Georgia or if it was the possibility that they were testing a Lyme vaccine on me while I was receiving those mystery infections. I did find it odd that two other soldiers were experiencing the identical things that I was. History has shown the Government and the Military conducting biological experiments on unknowing solders and civilians; some past examples were in Fort Detrick, MD or the Tuskegee incident.

Over the years since, my pain has gradually become worse and worse. In 2007, I ended up with the same rash again after cutting back some wild growth around my yard. I knew something wasn' t right when I got a severe flu in late spring. That was when I hit the wall, and the hellish migraine headaches began.

By mid-summer, they became completely incapacitating. Next, began the battery of tests that all Lyme suffers know so well. Since my main complaint was the migraines, I had numerous CT scans, MRI' s, and enough blood drawn to feed a fish tank full of leeches for a month. Subsequently, I found out that I had a 7-8mm cyst in my left frontal lobe of my brain and a lesion caused by the cyst. Finding the lesion, the doctors immediately thought MS. Resulting from that, I had to get a full head and torso MRI with contrast. That was so much fun; trying to lie perfectly still for an hour and a half on a plastic platform with all the comfort properties of a slab of concrete.

During that time period was also when the back pain worsened and the pain in my hips started. These pains continued to worsen over time and I began to experience serious cognitive problems.

By August 2007, I was experiencing numbness and weakness in my limbs. I started to feel as though I was walking with lead boots on. By early September, I started having trouble speaking. I was stuttering, forgetting words, or I occasionally I would try to speak to only form grunts and sounds. That trend continued until September 11th. That' s when I hit the wall. During the course of all this I had seen 2 different family doctors and a neurologist. They threw everything they could think of at my migraines. They had me on narcotics like Percocet, anti-depressants like Lexapro, anti-convulsants such as Topamax, to name a few but all which did squat.

So I awoke on the 11th barely unable to walk. By the time my wife came home from work, I couldn' t move my legs and my arms were very weak. My wife and I started to worry. Not long after, my tongue became completely paralyzed and to top it all off I had such an agonizing migraine that I couldn' t even open my eyes, because the lights felt like red hot pokers being driven into my head, through my eyes, by a sledge hammer. My wife called the neurologist and she said to get me to the hospital immediately. Arriving at the ER, they immediately wheeled me took to a bed. While I don' t remember what was happening to me, it appeared that my nervous system was shutting down. I had no reflexes or feeling when they were poking needles in my feet and other body parts. My right side of my face was drooping as well. They ran me through numerous types of neurological tests, which I had failed on all of them.

I was quickly admitted to the Neuro-Science Unit and they put me on IV Morphine. Occasionally, I would have small time frames where I try to communicate and recognize what was going on. Eventually between the pain, my cognitive problems and the morphine haze, I didn' t even know where I was. At one point when I was passed out, all I really remember was an excruciating pain in my chest and head. I felt tangled up, so I ripped at whatever was holding me back. Somebody tried talking to me, but I couldn' t understand the words. All I could manage to get out was that, " I had to get home" . From what I was told, I then tried to walk out. In my disoriented state, I was told that I then raged and they had to call a " code orange" , and it took 2 security guards and 3 orderlies to get me down. I vaguely remember hearing, " He' s seizing" . A few minutes later, I kind of came to and started to be aware of my surroundings again, but was in unbearable pain. I couldn' t stop screaming. I looked at the first guy I saw in a lab coat that was there and said, " Can' t you knock me out so I don' t have to feel this anymore!" That' s about the last thing I remember till the 15th of September.

I remember opening my eyes and seeing my family look kind of surprised. I kept nodding on and off till another doctor came in. This guy has all the bed side manor of a grizzly bear that someone kicked until it awoke out of hibernation. He said, " We' ve run every test known to man and we couldn' t find anything" . My wife insisted that they continue to look for what caused this when she saw that my discharge sheet listing my diagnosis as migraines and bipolar. She became livid and said there was no way this was due to that. He said," Well, we have already run $250,000 dollars worth of testing, and we don' t feel he is going to die right now so we are just going to let him follow up with his doctors. A hospital is just meant to stabilize you and get you good enough to follow up with your regular doctors" .

Upon speaking to my wife and my one on one care Intensive Care nurse a few moments later, I find out that they gave me a psych evaluation while I was unconscious and spoke to my wife about my time in the army and if I suffered from post traumatic stress due to my disoriented episode on trying to leave the hospital. The doctor blamed the aggression on bipolar and felt I was acting out for attention. (I now know that Lyme spirochetes in the brain can lead to aggression outbursts like that, so it was not a wonder it happened since I have a 7mm cyst of spirochetes in my brain).

While attempting to leave the hospital, when I tried to walk, I noticed that my right leg from the knee down was numb. We called back the doctor and he told me it was because I was just in bed for so long and insinuated that I was overreacting, told me I would be fine and then swiftly sent a nurse in to discharge me.

My leg never got better and I couldn't bear any weight on it. So that left me walking with a cane. I asked what happened while I was out for those 4 days. My wife told me that at one point they didn't know if I was going to make it. They had brought specialists in from all over and they couldn't explain what was going on. She brought me up to speed on everything that happened. To which I replied," See you can't get rid of me that easily. I'm too dumb to die."

Well after that whole fiasco, they put me

on IV medication for the migraines. That actually did help for quite a few months. After the migraines subsided, I was free to feel the pain in the rest of my body. It hurts very badly every day. There are many days that it restricts my activity. Then there's some that leave me completely incapacitated.

While trying to come to grips with my lack of mobility from trying to use the cane, severe muscle weakness, fatigue and the ever maddening pain, left me depressed to the point of contemplating suicide. This disease completely robbed me of my life. Here I am, barely play with my kids, no longer practicing martial arts, or working on cars. I'll never be able to run a 6 minute mile again or take my son for a nature hike. It's amazing how much you take for granted until it gets ripped away from you suddenly.

I managed to adapt somehow. I saw many new specialists, none which were able to help me or explain what was going on. One did feel that I had a stroke (We did find out later I did in fact have a Lyme inducted stroke). Then with the help of a Rheumatologist, I managed to lessen the pain. Even with lots of strong medications, it never goes away. The funny thing is that one of the first questions he asked me was, " Were you tested for Lyme disease?" I said that I thought I they had when they did my spinal tap, though he didn' t see it in my hospital reports. So just to be safe, he gave me the standard ELISA test. Well if you know anything about Lyme testing, then you know that even if you have a strong infection, your chances of getting an accurate test result are only about 50%. So of course I came back negative. Go figure.

Well then I really got sucker punched. My one bastion of strength and sanity, my wife, started to get sick. Our symptoms were so similar yet different. We both had the severe exhaustion and the cognitive problems. She ended up with seizures and heart palpitations though. In the end, the stress of her being sick made me worse, and stress of my condition made her worse. It was the most vicious of cycles. Finally I bucked up and just stopped talking about how bad I was really feeling. As time passed our symptoms slowly started to stabilize until a few months later my pain became so severe it caused a psychotic break. I ended up being a catatonic automaton. I went around completing daily chores, but unaware of anyone around me and unable to communicate with them or even be able to stop my movements. My wife had called 911. The ambulance took me to the ER when I collapsed to the ground, clutching my head while screaming. When I arrived, they pumped me full of enough pain meds to dope up a small town. After they broke through the pain, I finally came back to my right mind and was left to go home. One good thing did come of that though. My family doctor finally sent me to pain management.

There I was put on Lyrica and a Fentanyl pain patch. That helped me get past the pain in my back and hips long enough to work out my leg. Through that I was able to get enough strength, feeling, and control of walking without my cane on most days. I still have a limp but at least but at least I can get around on my own power most days. I've still never gotten full feeling back in my leg and foot.

It's almost about this time that my wife started seeing a Lyme Doctor near Philly. She seemed to like him and recommended that I try to go to him and see what we could find out.

The day I went down to see him, he seemed very short and annoyed. I thought he was just having a bad day. So his nurse drew blood for and ELISA test. This in case you don' t know, tests for if your body is trying to fight Lyme disease. Well that test is about useless because Lyme is a type of spirochete. It burrows into tissue and hides from the immune system. He also ran a Western Blot test. That test is more useful and can even help narrow down the strain of Lyme a person has.

Well the ELISA test came back negative, but the Western Blot had a very interesting result. I had a very high and positive Band 23, 30 and 31 (31 is also the band that the Lyme vaccine triggers, which made me question about those injections back in the Army since I never received the vaccine on my own). These are very specific Lyme bands. I also was just under positive for a Band 41 by a few points. To this, the doctor told me that I was definitely exposed to Lyme and he was certain it had been going on for quite some time. He then explained that if I had different insurance and that if my Medicare would have started already since I was now on Social Security Disability, he could have had me on IV therapy that same day. So instead he put me on oral Zithromax. Within a few days, I started to have a really bad

Herx reaction.

When my next appointment came up, I spoke with the doctor and explained to him what my herx reaction was like. I also had to talk to him about losing my Blue Cross insurance and wanted to see if my wife and I could work out a payment plan. I guess it shouldn' t have surprised me, when he had the gall to look at me and say, " Well you' re just allergic to the antibiotics and since we have tested you multiple times and your ELISA came back negative; I don' t think you have Lyme and I am discharging you." I was so stunned that I couldn' t even try to argue as he rudely spun around and walked out. What I would have said was something to the effect of- what multiple tests?? And what the hell happened to you saying I was definitely exposed to and positive for Lyme disease last time I was here??? Alas, I will never get that chance.

Of course after my wife and I both got sick, she did a hell of a lot of research. That' s how we both learned how hard it was to get proper treatment for Lyme. So what happened to me was anything but uncommon. From all the research my wife did and all the stories we' ve heard; it seemed like treating

Lyme was like the back alley abortions of the 1950's. Except it's the government and insurance companies keeping us from getting the treatment we need. I know about the insurance companies first hand from the inside and the medical provider side. I at one time was a licensed insurance provider. When I saw how many of my customers got screwed on legit claims, I got out quickly. I had too much of a conscience for that. I was also a medic in the Army. After that, I went to work in a hospital.

The insurance companies all but run the hospitals. If you don't have insurance, all you get is stabilized and then a boot to the door. That seems to be a problem with our government in general. Instead of a democracy where everyone gets a say; we've become capitalistic so the big businesses get almost all the say. When a state attorney general finds that the people of the IDSA panel have too many conflicts of interest to impartially make a decision, you know they're messed up. As a matter of fact, I beg you to check the facts about the IDSA panel members and see for yourself. People's lives are being destroyed because their " people" (and I use that term lightly) wanna make more money. It sickens me to see the effect that this has had on my family and the lives of the people we have come to know through this ordeal.

No matter how hard I try, I can't give my kids a "normal" life. I can't go out and run around with them like most normal dad's can do. If I try tom after a few minutes, I end up so exhausted that all I want to so is come in and lie down and sleep. It's not just my quality of life that concerns me. It's the fact that I know my death clock has started ticking. I know that without proper treatment, it's only a matter of time until my condition progresses into something more serious.

Through the research of Dr. Alan MacDonald, we've learned that a majority of Alzheimer's, MS, and ALS patients were actually positive for Lyme. He found this by taking and examining brain tissue samples. I could go on all day about the different research that is being or had been done. All I really want is for a truly impartial panel to look over all of Dr. MacDonald's and Dr. Burrascano's work and all the research done in Texas. That way, a truly informed set of treatment guidelines can be set. If that can be accomplished, then my hope is that I might have a chance of being cured. Then I could also come off of Social Security Disability and be a productive member of society again. That and I also hope I can try to be a better dad to my kids.

So if you, the reader, can do nothing else, please check in with your federal law makers to see how the re-evaluation of the guidelines are going especially since they are subject to mandatory review in Summer of 2009 because of Attorney General Blumenthal's investigation. Even if it's for no other reason then to get me and the thousands of others like me off of federal disability. Our law makers need to know how many of us are interested in this and that we are appalled by the IDSA's treatment of the subject thus far. Please help us to live normal lives again.

Trish
Massachusetts

My name is Trish. I currently live in Sturbridge
MA for 11 years. Born in PA lived in Colorado
and then NJ. I was told for over 10 years I had
Fibromyalgia. One morning I woke up to get my
two boys off to school when one of them said to
me MOM you don't look so good. I was very
dizzy and my head felt odd. I assured them I
would be fine and was going to rest once they
left.

Within a short time the top of my head went
numb and something inside said "something is
really really wrong." I called my husband at
work who assured me he's too busy and I would
be fine. I freaked out and was completely
hysterical. Something in my body was telling

me something is wrong. I hung up the phone and sat in a chair where I became completely unable to feel or move anything. The paralysis worked its way down from my head to my feet. Phone rang. It was right next to me but I couldn't move to answer it. Thankfully my husband called an ambulance from work and they came. When they went to pick me up my body was limp! I was terrified. It took 3 of them to get me on the stretcher. On the way the paramedic made eye contact with me and realized I was in there. He started asking me questions and I was to blink 1 for yes twice for no.

Upon arrival at our horrible local hospital I was wheeled into a room where they looked me over. Tried to look in my mouth but my jaw would not open. The thoughts were running thru my brain so fast but nothing was coming out of my mouth! This doctor sat me up on the gurney and just let me go...I quickly wilted over the side rail and was caught by a nurse as the room gasped!

To my disbelief the doctor said " leave her alone, when she wants to talk she will!" My heart was pounding I couldn't believe what I just heard. I began praying to all of my angels and I didn't stop. After some time nurses came in and acted as if they had been with me for some time followed by my husband. I guess they couldn't let him know that I have been lying there alone the whole time.

At this point I just knew I was having a stroke and trying to make myself accept the fact that "I'm going to die in this crummy hospital."
I began to calm myself and gave it up to God. At some point, seemed like hours I finally got a CAT scan. I was wheeled down with only one nurse at which time I could feel my insides going numb and couldn't breathe well. I began to panic and blink until the nurse realized I was having trouble. They sat the bed up a bit and it helped but they had to lay me flat again to get the Scan. With one nurse and my dead wait I nearly rolled off the table and onto the floor. I was helpless and at the mercy of a nurse who at this point was trying not to cry and assuring me she would be right back and sit me up again to help me breath.

Anyway nothing much was happening at this hospital and my husband asked that I get transferred immediately to Boston. This took hours.
I arrived at Brigham and Women's hospital at

night when I saw a nurse look down my way and said "oh my god look at that facial droop." I'm thinking, poor bastard someone is in trouble. To my horror they came running to me!

I remember them asking me questions but I couldn't answer and my husband wasn't there yet. My mind raced with thoughts that I would be this way forever if I didn't die. My boys, what will they do without me? I prayed and was taken to an MRI machine. I laid in this machine for over an hour as it broke down several times. Then I hear the technician yelling at someone that this machine isn't even calibrated for adult stroke and as I laid there paralyzed and suffering from claustrophobia I couldn't believe what I just heard. These are the films that they will have to look at to see where my brain has failed!

After films were read they said they didn't see a stroke and they were calling psych down for consult. What? I thought to my self and my husband out loud. Beside the fact the previous hospital only sent a note saying I came to the ER complaining of a sore throat!
Unbelievable. No mention of an ambulance, paralysis, inability to communicate.

Psych came they put me in a small room and told my husband to go home. Long drive. By this time I could move some of my right side and my head as well as yes and no words. It was early in the morning hours when the consult took place but I managed to communicate. Next I have psych doctor arguing with ER doctor that this is not a psychological problem, this is absolutely physical and I needed to be admitted. Without a picture of a stroke ER doctor refused and psych couldn't do anything for me. I was then told to get out of the bed they needed it! My husband was asked to leave. I was completely paralyzed on my left side still from the face down. And I'm being told to get up and get out. I couldn't believe it. Had them call my husband who came back hours later because of the long commute, long night etc.

I was paralyzed, could barely speak and I was being wheeled out of the hospital being told this is stress go home! Both in shock we were to 5 different hospitals in the Boston area in 6 days. Apparently once in the central system they look up your name and whatever the first hospital diagnosis was that was it. Except for the fact that I would get a psych consult at every hospital after a wait of about 12 hrs. it always ended up with psych arguing with ER doctors that this is real not psych and I needed to be admitted.

I also was suffering constant and severe migraines all day everyday as I would once again be wheeled out to the street being told nothing we can do. I was sent home and started working with physical therapists who took one look at me and left. She was told I had a mini stroke and one look at me she said "I can't work on a stroke patient." The next therapist came the next day took my blood pressure that was thru the roof and immediately called my PCP. They without a care said, " oh yes we will see her tomorrow she's fine" . My Physical therapist was so angry and asked to talk to the doctor and tried her best. She could not work with me that day it was too dangerous in my condition. We tried desperately to get help for at least migraines which were still all day everyday. After a few weeks I was given something but didn't work. My husband and I figured we better try a therapist because I was really sick and in severe pain from migraines. The therapist took one look and said "it's Lyme and you are the worst I've seen⋯" We were shocked, confused.

Anyway she was the first to find a Lyme doctor for us though out of state we had to pay out of pocket but what choice did we have.
Doctor immediately started me on antibiotics and within days I could speak and think better, find my words and physically I began to be able to move my left side. It took months to get my left side where it is today. Because I was diagnosed so long after the paralysis began the lyme doctors believe it's permanent now. They also believe I may have had a stroke that wasn't caught because most lyme paralysis goes away with treatment. I am now with a lyme neurologist in Norwalk CT who we also have to pay out of pocket because we were unable to find a LLMD in MA. It has cost us tens of thousands in tests, meds, and visits. What choice does a family have when hospitals send people away in America, unable to walk, talk, think?

My lyme has now involved my GI system and my pancreas barely works. 35% function if I'm lucky. This from years of having pancreatitis that went completely undiagnosed for years. I also have Celiacs disease and am on a Gluten free diet and Gastropharesis which is the paralysis of the Vegas nerve which causes slow stomach emptying. This is a disease that is directly related to lyme. My GI has never seen this with Celiacs and urges me to continue the treatment for lyme but avoid oral drugs as I can't tolerate them.

I have waited months now for an appeal. Approval for a PICC line came two weeks ago but the insurance company won't allow any meds to go into it. I have the line but no meds. How does this make sense? I have my PCP, LLMD and Pancreatic specialist all helping but I wait and whither away. I have been bed ridden for about a month and a half now and all I can do is pray the insurance understands the extenuating circumstances and listens to the professionals to keep my CNS Lyme from killing me. And it is!
Thanks you for listening. I pray for those who have lost their lives from this epidemic.

Renee Blaker
Doylestown, Pennsylvania

I was born and raised in Huntington, Long Island, New York on the North shore.
I went to college in Washington, D.C.
I spent a few years living in Philadelphia prior to moving to
Doylestown, Central Bucks County, PA where I still reside.
I'm a wife, mother of 2, speech-language pathologist, artist, gardener, reader, cook, movie lover, dog lover, good friend, volunteer, just living my life.

I went undiagnosed 10-15 years according to
my Lyme doctor.
I have traveled to many places both in and
outside of the U.S.
I certainly have been bitten by ticks in my
lifetime, but never before knew there was such
a thing as a deer tick, or that there was a
disease caused by a tick bite.

One day, clumps of my hair begin to fall out.
Sure, I was under stress, but what was
happening to me felt surreal.
You have Alopecia Universalis, a doctor said.
That means total hair loss, but why.
Every hair on my head and body eventually was
lost.
I was completely bald. Even the tiny hairs in my
nose disappeared.
At first I was in denial. Then, I was frightened.
I felt like a freak.
I was even told all my finger nails might fall off.

No one ever said Lyme, or anything else for
that matter.
There were other problems. One doctor told me
I had Fibromyalgia.
So. Nothing explained my trouble in focusing
my eyes, one eye seeing higher than the other.
I had terrible migraine headaches, neck and

back pain, often debilitating, unusual fatigue,
one knee with some fluid and slightly swollen.

I had a circular quarter size rash on my lower
torso that just appeared and stayed for years.
Other random rashes, nothing very significant.
My right eyelid drooped. No one took notice.
You must be tired. I saw my nose in front of my
eyes constantly. Just ignore it, I was told.

I had a lot of floaters. Panic and anxiety
attacks took over my life, although I hid them
from everyone except my husband for years.
Irritable Bowel Syndrome caused me to be
afraid to leave the house for fear of an attack.
I was claustrophobic, agoraphobic, and terrified
all of the time, but I continued to live my life as
if all was well. My reflexes were hyperactive. I
startled so easily I could hardly be a passenger
in a car.
I was afraid of being found out. What could
possibly be wrong with me?
I must be crazy.

The list of symptoms kept getting longer.
I became depressed.
I thought I had an ulcer.
Then there was the excrutiatingly painful
bladder irritation and

incontinence.

Also the high frequency hearing loss and tinnitus (ringing in ear), however my hearing is just fine now.

Tests were always negative or inconclusive.

Part of me was relieved.

But, I knew in my heart, body, soul and mind that the doctors were missing something.

I WAS really sick.

I still worked 12 hour days in hospitals and my private practice,

Helping other people who couldn't speak and/or swallow.

I started misspelling words, had difficulty thinking; I forgot things easily.

I had difficulty expressing myself.

I would talk around a topic, unable to get to my point.

Again, the symptoms were subtle, but very real and frightening to me.

I choked on my food.

I lost my active gag reflex.

It took me hours to write chart notes.

Time escaped me.

I got lost trying to get to familiar places.

I dropped things.

I became confused; I had difficulty processing

information.

I had weird blood test results which were
dismissed.

Surely, I was losing my mind.

But again I thought I should feel relieved.

Nothing horrible was wrong with me.
Maybe I had M.S.? I felt like I did. I would
experience transient numbness and pain,
tingling sensations and difficulty with balance.
I would walk into walls.
Trying to sit down on a toilet seat, I would
partially miss the seat.
I always felt like I was falling.

Then one day a friend dragged me to his
doctor/friend.
He thought for sure this man would figure out
why I lost my hair and help me. I saw every
specialist I could think of in addition to this
doctor.
My friend's doctor didn't make the Lyme
diagnosis at first.
Eventually, he tested me for Lyme. He had
Lyme himself. If he hadn't known first hand the
ugliness of this disease, I think he would also
have missed the diagnosis. He treated me with
respect and took me seriously. He spent time
listening to my story. The diagnosis was based

on clinical findings and test results and my case history. He took a very thorough case history. He jumped to no conclusions.

Then, in the middle of everything, still trying to work, as I was the main bread winner in our family, I fell asleep in front of one of my patients. He had had a stroke and could not talk. Apparently, I awakened shortly and thought his session was over after only 15 minutes had elapsed.
He was the last patient I treated for 7 years.

IV antibiotics on and off for 7 years, and a variety of other therapies consumed me. Many pharmaceuticals, herbs, acupuncture, massage, hyperbaric oxygen, you name it, I may have tried it.
I got so sick that eventually I walked with a cane, could not drive a car, often could not lift my head off my pillow. I remember being wheeled through an airport on my way to the Mayo Clinic; being wheeled through an art museum trying to feel something other than sickness. How humbling these chair rides were for me!

The headaches were excruciating.
I thought my head was going to explode.

I fell. I held onto walls trying to make my way
down a hall.
I' ve had 2 picc lines, then 2 central lines.
Infections; lines that threatened my life.

I was hospitalized on many occasions with
"fevers of unknown origin".
My temperature would hit 103, and then go
down.
Other times it was 95 or 96 instead of my
always normal temp of 98.6.
Finally, I was hospitalized with Meningitis for 2
weeks.
I had PET scans, and SPEC scans and spinal
taps more than once.
I wanted to die,
But, I didn't.
I slowly got better.
I relapsed several times until I stabilized.

2 brilliant doctors kept me alive, as well as the
support of my family, friends, & other health
care professionals.
Now, I can work about 4 hours at a time, a few
times a week, instead of 10 −12 hours most
days like I did for so many years.

At 62, financially, I am in debt, instead of
thinking about cutting back my hours voluntarily

to spend more time with family and friends.
I have 3 grandchildren (and 1 on the way), who
I cherish with all my heart. (Lilah, Charlie &
Jack).
I have 2 grown married children and their
respective spouses who I am so very proud of
and love dearly.
I have a supportive husband who has stayed
with me for over 40 years through Lyme rages
and Lyme paranoia and all the IV care he
administered when our insurance company
failed us despite outrageous premiums.

Life goes on....I am glad I kept the fight going
for so many reasons. And it was a fight. My
heart goes out to all the Lyme patients who
must fight the fight and go through hell and
back in order to heal.

I believe most of us will make it if we find the
strength, support, medical care, and financial
care we so rightly deserve for this illness, just
like any other mental or physical illness.

My immune system has been attacked.
Now I am in the process of putting back the
pieces.
I try not to look back too often.
It's hard for me to remember all the years of

agony. It often feels like it happened to someone else who is now a small part of me, but still needs nurturing all the same. I guess that is my mind's way of protecting me.
The residual neurological difficulties remain the most exasperating to me, but I am grateful for the extent of my recovery.
I thought I would surely die.

So hang in there....if you possibly can.
You are worth the fight.

Terra Hall
Salisbury, Maryland

My name is Terra. I was born in May of 1987
and live in a small town in Maryland and my life
revolves around 22 + pills a day. I have
Chronic Lyme Disease the great mimicker. At
the age of 8 I came down sick thought I had the
flu but it was not flu season so my doctor ran
some tests and I was told I had Lyme disease. I
went through a 2 week treatment of antibiotics
and never thought of it again there was no need
Lyme disease is nothing to worry about right?
WRONG! I was one of the most at risk people in
my area. I lived out in the woods and would get
bit by 1 to 50 ticks a day; and I am not

stretching it. I lived in the most of rural areas had a lot of pets and spent most of my time wondering the forest and the fields around my house. Yet I had never heard of Lyme and once I did it was just " oh you get that rash thing."

In the upcoming years I sprained this, pulled that, twisted another joint, I spent 50% of every school year on crutches. It did not stop me though I was a cheerleader, I played soft ball, I ran every day, and weight lifted 3 times a week, basically lived an extremely active life style. I always got tired faster then most and was sick on and off more then most, a lot more then most but my parents never thought much of it I was a young person we get hurt we get sick never thought it could be more. In the upcoming years we realized I had dyslexia··· kind of out of no wear. Once again it did not slow me down; I still participated in public speaking contests, and regularly won essay contests I just had to work a little harder then most. Little did we know that those things were signs of so much more going on just a tip of the ice burg below the ocean level.

When I was 14 things started to get a little more interesting. In the 3 years following my freshman year in high school I was told I had mono 3 times, but wait you can only get that once. The flu 2 times when it was not flu season. My sophomore year I missed around 42 days of school because of an obscure injury or illness, and every following year that number increased. It got so bad my jr. and sr. year that I had attendance meetings for missing to many days of school and risked repeating those school years but I was fortunate enough to have a faculty at my school that understood, I was not just skipping school I was sick··· A LOT!

The summer before my Sr. year things rely took a turn down the wrong road. I went out with friends dancing and hurt my back. Not a big surprise but what scared us was I did not get better in fact I could not get out of bed. My Dr. of the previous 2 years said " I' ve had enough." He ran a long list of tests and what did you know··· I HAD LYME DISEASE and on top of that Babesia also. The following 3 months were hell on my body. I started a 3 month treatment of Doxy. I lost most of my eye sight could not walk could only sleep and then could not sleep. I was on 15 pills a day and missed the rest of that summer and the majority of my first 2 months of my Sr. year. It got so bad the local Dr.' s looked at my mom and told her " I am not sure if Terra will see graduation." My health slowly seamed to level out and what did not get better I got used to. I ended my round of medication and we thought I was ok. I had accepted I would never play sports again, and I would probably never have the life I once had. Little did I know the long term problems I would have and once again what I had just went through had done little to nothing to improve my health.

Since I was 17 I have undergone 4 or 5 forms of treatments each less successful then the last. I have seen 10 plus doctors and have been told I was faking it, I still had mono, I just got the flue a lot, I was a klutz, and the best " your just depressed because you don' t have a boyfriend." I have tested positive, and inconclusive a number of times for Lyme disease there is no doubt in any ones mind at this point I have and still suffer from Lyme disease.

As of August 2008 I have come down ill again more so then ever before just some of my symptoms include racing heart, dizziness, fevers, difficulty concentrating, forgetfulness, extreme joint pain, muscle pain, muscle tremors, memory loss, brain fog, anxiety attacks, blurry eye sight, insomnia, migraines, numbness of extremities, light sensitivity, blood pressure and sugar changes and fatigue. I am once again undergoing treatment this time my body is shutting down more so then it did in my first hard core treatment. I have had to leave college where I was a double major in social work and sociology. I have also left my full time job on medical leave as this disease destroys every part of my life.

I spend days on end in bed only to get up to do the most basic of things in the house. I spend most of my time that I should be resting, fighting insurance companies to please allow me to undergo the treatment I need. Because most of the oral antibiotics have proven ineffective on me because the Lyme disease in my body is so strong. Those oral antibiotics that are strong enough took a strong toll on my stomach and organs. So I am fighting to receive an IV treatment with hopes that I can attack this disease hard. I want nothing more then to be back in college go back to work.

I want to be 21 to go out to have fun. It is not fair that because my disease is swept under a political rug I have to suffer. I WANT MY LIFE BACK! I by no means am asking to be perfectly healthy have I just wanted with all my heart to live a stable life not waiting every moment for a rush trip to the closest hospital.

I fight now just as hard as I did years ago then it was to play sports and write clearly. Now it is to stay out of the hospital keep my eye sight and maybe just maybe get back to school and work. If my disease had been treated truly back then I may not be fighting this fight. How ever I can not blame my doctors they knew little of this disease they did what they thought was best. That's why I have fought with my doctors to get more information to the local medical community and have at last found a truly lyme literate doctor to assist in my treatment. I can only wonder the what if it had been caught and treated properly so many years ago.

Lyme disease has taken years of my life. I don't want to lose another.

Update from the author: Terra has been on IV treatments. Her disability insurance denied her claim and told her she was healthy enough to maintain working, so she is back to work fighting as hard as she can. While it is unheard of to try to do this, she has no other choice because Lyme Disease is not currently listed as a disability.

Sara Keeton
Frederick, Maryland

My name is Sara Keeton and I would like to tell
my story of battling Lyme disease and tick-
born illnesses. I have been suffering with Lyme
disease, Babesiosis, and potentially Bartonella
for the past seven years. Lyme disease has
altered every facet of my life. This disease
has, in countless instances, claimed my self-
confidence, my mind, my body, my
independence, my friends, and my relationships.

I want to caution anyone who is ill and
searching for an explanation to be secure in

your knowledge of yourself. Many medical professionals have a unique ability to convince us that they know who we are and how we feel better than ourselves. I am not an exception, and although I always held that there must be "something" more than a profoundly quick onset of psychological illness amiss with me, I was convinced for some time that I must be exacerbating the problem in my mind and creating physical manifestations and symptoms that did not have a physical basis. I should not have had to doubt what I knew to be true. I should not have had to question whether my pain and my suffering were real. No one should have to fight to prove their mental stability on top of fighting chronic infection. Doing so puts an undue strain on our bodies and minds— valuable energy that we should be using to fight this disease.

One of the aspects of my Lyme journey that strikes me as blatantly illogical is the notion by medical professionals that "we do not know what is wrong with you; therefore, there is nothing wrong with you." This is simply faulty logic. It assumes that the medical professional must know everything there is to know about medical conditions and lack no knowledge on them. To draw such a conclusion necessitates ruling out every other possibility. In order to rule out every possibility one would have to not only be aware of every possibility, but be aware of all measures needed to rule them out. Similarly, drawing the conclusion that "we cannot find anything physically wrong with you; therefore, the basis for your complaints must be psychological" seems dangerously presumptuous. Again, to be able to draw such a conclusion, one would need to be aware of every physical ailment, every usual and unusual manifestation of every physical ailment, every test to perform to concretely rule out the physical basis of every ailment, and such tests would have to be foolproof. It is incredible to me that so many medical professionals not only draw the conclusions that nothing is wrong with us or that what is wrong with us lies in a psychological malfunction, but that they are able to make that determination, often, in a matter of minutes! Impressive! There is an old adage that states, " The more you know, the more you know you don' t know." Perhaps our physicians should heed this when they are confused by our myriad of seemingly unrelated symptoms. If they were wise, they would be aware of their potential lack of understanding and would admit that it' s possible they just don' t have the answers. This should not be a substitution for continual education on behalf of our physicians, and a drive to find the answers to difficult questions

I would advise anyone who has seen or is seeing a doctor who has told you that your ailments are not real, or that they are in your head, to fire them immediately. Erase every ounce of medical advice that they may have given you prior to drawing such an egregious conclusion. It is one thing to not know what is causing a patient' s symptoms. If such is the case, you should expect the medical professional to tell you exactly that, "I'm sorry, Mr. or Ms. _____, I do not know what is causing your symptoms." If they are decent, they will then attempt to find the cause your symptoms. They should take into account that there are illnesses (such as Lyme disease), for which diagnoses are based on clinical information. In such cases, they should take pains to listen to your complete medical history, and to rule out conditions as much as they would try to rule a particular condition in. If they still cannot find the cause of your symptoms, perhaps they should consult with other physicians or do some more reading. If they are completely stumped and want you to get well, they might refer you to someone who they think can better help you. They should never tell you that there is nothing wrong, or assume that your symptoms must have a psychological basis simply because a physical basis has not yet been found.

Personally, I have never suffered with any form of psychological illness. I had a normal, healthy childhood with two loving parents, a supportive sister, and caring friends. I was never an outcast, did not suffer any early childhood trauma, and even made it through my teen years relatively unscathed and without suffering the common woes of depression, or struggle for self-actualization. At times, I created rifts in certain relationships with friends because I had a difficult time relating to people who did go through the teenage tussle. I was almost arrogantly grounded and self-aware.

I didn't visit the doctor very often growing up. Having been a competitive athlete until the age of seventeen, my immune system was strong and I remained happy, healthy, and active. When I started to have frequent complains of poor health, despite my reluctance, I began visiting doctors frequently. I had shown no prior pattern of hypochondria, or an inability to deal with the aches and pains of daily living. In fact, I was quite the opposite-exhibiting a high tolerance for pain and physical discomfort. So, it came as a big surprise when my family jumped on the bandwagon of doubting that my complaints had a physical origin and even in some cases thinking that I was crying out for attention. I was very confused since I had no history of seeking attention in unusual ways or even requiring much attention in general. I continue to be baffled by the many who share in my experience, having had their family and friends turn on them in their greatest time of need. How sad, when what we have needed most (aside from some basic humility and more knowledge on this disease by the medical community), has been an advocate who would believe us in the face of those who were drawing such hasty and inaccurate assumptions. Those of us who are lucky will even go on to test positive for Lyme disease and/or other tick-born infection despite the horribly inaccurate tests used to detect these illnesses. It is dumbfounding that even with positive test results (which I will stress, are in no way necessary for diagnosis of Lyme); our family and friends still question our motives for seeking treatment by Lyme literate doctors after watching our struggle to obtain a proper diagnosis. They have witnessed us bounce from doctor to doctor, only to be sent home with no answers. They are aware that we've been told we're mentally unwell, or told we're fine, or misdiagnosed. They know we've been given the wrong answers, so why would they trust the same doctors with our treatment protocol when we do test positive and do get diagnosed?

I wish someone could explain this to me:

On one hand we have a doctor tell us that we're mentally unwell or perfectly well, we know that's not the case, meanwhile we exhibit the clinical symptoms of lyme disease. On the other hand, we meet a Lyme Literate doctor, we're diagnosed with lyme disease based on those clinical symptoms, and then some of us go on and test positive for the disease. The first doctor contends that lyme disease is curable after a few weeks of antibiotic treatment. The lyme literate doctor contends that it can be a very long road, requiring high doses of antibiotics over months or even years. If you were one of us, who would you go to for your treatment? The doctor who called you crazy, says there was nothing wrong with you, failed to test you for Lyme or if he did failed to explain that the tests are inaccurate, and sent you home? Or, would you return to the doctor who recognized your symptomology, diagnosed you, tested you using more accurate testing, understood the necessity of clinical date for diagnosis, and all the while believed the reality of your suffering? Coincidentally, the doctor in the latter example seems to repeatedly achieve a significantly higher patient response to the treatment he or she prescribes.

Victims of Lyme disease and tick-born-illnesses do not want to sit in bed all day. It's not fun for us to miss out on activities and events. It isn't laziness or lack of ambition that prevents us from working. Most of us would love to work. Most of us need the money to get well. Most of us like human interaction. Most of us do not take pleasure in not being able to contribute to our families and our communities. We don't like spending thousands of dollars out-of-pocket to see doctor after doctor. We don't like taking our time and energy to drive to the doctor's office, be examined, repeat our story over and over, often only to be dismissed and told to go home and lead a normal life. We do not enjoy being looked at by our friends and families like we have a mental deficiency. We don't like having to borrow money and resources from the people we love. We don't like going to the emergency room in the middle of the night time and again only to be given a dose of pain medication and some IV fluids and sent home. Please believe me when I say that this isn't fun for us. If we were merely crying for attention, don't you think we would find a more beneficial way to get attention without bankrupting ourselves and our loved-ones, destroying our credibility in the eyes of the people we respect the most, severing our relationships, and missing out on the activities we have always enjoyed?

We know that many of us "look well." We know that some of us have "good days" when we can get out of bed and participate in activities. We know that it seems odd to have days when we are completely debilitated and days when we can carry on normal conversation and accomplish household chores. It's an odd disease. We know that it is confusing when our symptoms are varied and changing and contingent on the day. We know how our claims come across to people who think that searching the internet or having a conversation with their non Lyme literate physician is going to give them accurate answers. They approach us and tell us that the internet says that Lyme is curable with a few weeks of antibiotics, or that their physician told them there's is no such thing as chronic Lyme disease. They think that they have put in the effort to do their own research and that they've validated what they have thought all along--we're crazy. Even when we try to educate them or repeat to them that the information they've obtained is false, they don't believe us. Let me ask those people--what advantage would there be in choosing to align ourselves with a school of thought that is considered outrageous by the vast majority of the medical community? Don't you think we would be elated if Lyme disease and co-infections were, in fact, cured with a few weeks of antibiotics? Do you think it's enjoyable to take handfuls of expensive medications daily, sometimes intravenously, which carry risks and unpleasant side-effects? Do you think we would do it if it didn't work? For those of us who suffer from chronic Lyme infections, we know that often doing so is the only thing that works. We have been researching this disease for years, but more importantly, we have been suffering from this disease for years. In my case, I have been sifting through the masses of contradictory information in the literature and spoken by medical professionals for five years now. The antibiotics work. The Lyme Literate Medical Doctors are right. Chronic Lyme disease exists. Chronic Lyme disease is debilitating and disabling. Lyme patients are not crazy, they're sick. Please, if you love someone who suffers from Lyme disease, do not work against us. Be the one person who will stand up for us, trust in us, and help us. Thank you for listening to my story.

Epilogue

❧❧

In parting, I wanted to share with you the very educated testimony of Dr. Joseph Burrascano, Jr. MD. Dr. Burrascano is a hero in the Lyme community. The very well respected Lyme literate doctor gave the following testimony below to the Senate Hearing Committee on Lyme Disease. Dr. Burrascano stated in his testimony that he feared repercussions for speaking out at this hearing. His fears were founded, unfortunately. Several years down the road, he was investigated by the Office Of Professional Medical Conduct in N.Y. Many activists and patients whose lives he saved helped to raise money for his legal expenses. He was finally vindicated and did not lose his license unlike several other doctors who have treated chronic Lyme and were punished for it. He has always been a leading spokesperson always speaking out to bring about change. The education and treatments provided by Dr. Burrascano have truly been an asset in our fight with Lyme. The world is a better place thanks to him.

The Lyme Disease Conspiracy
by Joseph J. Burrascano, Jr., M.D.
Reprinted from the Senate Committee Hearing on Lyme Disease
August 5, 1993

There is a core group of university-based Lyme disease researchers
and physicians whose opinions carry a great deal of weight.
Unfortunately many of them act unscientifically and unethically.
They adhere to outdated, self-serving views and attempt to
personally discredit those whose opinions differ from their own.
They exert strong ethically questionable influence on medical
journals, which enables them to publish and promote articles that
are badly flawed. They work with government agencies to bias the
agenda of consensus meetings, and have worked to exclude from
these meetings and scientific seminars those with alternate
opinions. They behave this way for reasons of personal or
professional gain, and are involved in obvious conflicts of interest.
This group promotes the idea that Lyme is a simple, rare illness
that is easy to avoid, difficult to acquire, simple to diagnose, and
easily treated and cured with 30 days or less of antibiotics.

The truth is that Lyme is the fastest growing infectious illness in
this country after AIDS, with a cost to society measured in the
billions of dollars. It can be acquired by anyone who goes
outdoors, very often goes undiagnosed for months, years, or
forever in some patients, and can render a patient chronically ill
and even totally disabled despite what this core group refers to as
"adequate" therapy. There have been deaths from Lyme disease.

They feel that when the patient fails to respond to their treatment
regimens it is because the patient developed what they named "the
post Lyme syndrome". They claim that this is not an infectious
problem, but a rheumatologic or arthritic malady due to activation
of the immune system.

The fact is, this cannot be related to any consistent abnormality
other than persistent infection. As further proof, vaccinated

animals whose immune system has been activated by Lyme have never developed this syndrome. On the other hand, there is proof that persistent infection can exist in these patients because the one month treatment did not eradicate the infection.

Indeed, many chronically ill patients, whom these physicians dismissed, have gone on to respond positively and even recover, when additional antibiotics are given.

It is interesting that these individuals who promote this so called "post-Lyme syndrome" as a form of arthritis, depend on funding from arthritis groups and agencies to earn their livelihood. Some of them are known to have received large consulting fees from insurance companies to advise them to curtail coverage for any antibiotic therapy beyond this arbitrary 30 day cutoff, even if the patient will suffer. This is despite the fact that additional therapy may be beneficial, and despite the fact that such practices never occur in treating other diseases.

Following the lead of this group of physicians, a few state health departments have even begun to investigate, in a very threatening way, physicians who have more liberal views on Lyme disease diagnosis and treatment than they do. Indeed, I must confess that I feel that I am taking a large personal risk here today by publicly stating these views, for fear that I may suffer some negative repercussions, despite the fact that many hundreds of physicians and many thousands of patients all over the world agree with what I am saying here. Because of this bias by this inner circle, Lyme disease is both under diagnosed and under treated, to the great detriment to many of our citizens. Let me address these points in more detail.

UNDERDIAGNOSIS
1. Under reporting: The current reporting criteria for Lyme are inadequate and miss an estimated 30 to 50% of patients. Some states curtailed their active surveillance programs and saw an artificial drop in reported cases of nearly 40%, leading the uninformed to believe incorrectly that the number of new cases of Lyme is on the decline. The reporting procedure is often so

cumbersome, many physicians never bother to report cases. Some physicians have found themselves the target of state health department investigators. Finally, to many physicians and government agents rely on the notoriously unreliable serologic blood test to confirm the diagnosis.

2. Poor Lyme disease diagnostic testing: It is very well-known that the serologic blood test for Lyme is insensitive, inaccurate, not standardized, and misses up to 40 percent of cases, yet many physicians, including many of those referred to above, and the senior staff at CDC and NIH, insist that if the blood test is negative, then the patient could not possibly have Lyme. This view is not supported by the facts. Lyme is diagnosed clinically, and can exist even when the blood test is negative.

The Rocky Mountain Lab of the NIH, which is the country's best government laboratory for Lyme research had developed an excellent diagnostic test for this illness nearly 4 years ago, yet further work on it has been stalled due to lack of funding. Incredibly, if not for private donations of just $5,000 from the non-profit National Lyme Disease Foundation headquartered in Connecticut, then this research would have had to be abandoned. An additional $30,000 was donated by this organization to allow them to continue other valuable projects relating to vaccine development and disease pathogenesis. Yet, many physicians believe that thousands of dollars of grant moneys awarded by the government to other, outside researchers is poorly directed, supporting work of low relevance and low priority to those sick with Lyme. In spite of this, their funding continues, and the Rocky Mountain Lab is still under funded.

3. The university and Government based Lyme establishment deny the existence of atypical presentations of Lyme and patients in this category are not being diagnosed or treated, and have no place to go for proper care.

RESULTS: Some Lyme patients have had to see, as many as 42 different physicians often over several years, and at tremendous cost, before being properly diagnosed. Unfortunately, the disease was left to progress during that time, and patients were left forever ill, for by that time, their illness was not able to be cured. Even more disturbing, these hard line physicians have tried to dismiss these patients as having "Lyme Hysteria" and tried to claim they all were suffering from psychiatric problems!

UNDERTREATMENT
1. Because the diagnosis is not being made, for reasons partly outlined above.
2. University based and government endorsed treatment protocols are empiric, insufficient, refer to studies involving inadequate animal models, and are ignorant of basic pharmacology. They are not based on honest systematic studies or on the results of newer information.
3. After short courses of treatment, patients with advanced disease rarely return to normal, yet many can be proven to still be infected and can often respond to further antibiotic therapy. Unfortunately, Lyme patients are being denied such therapy for political reasons and/or because insurance companies refuse to pay for longer treatment, upon the arbitrary and uninformed advice of these physicians, who are on the insurance company's payroll.
4. Long term studies on patients who were untreated or under-treated demonstrated the occurrence of severe illness more than a decade later, reminiscent of the findings of the notorious Tuskegee Study, in which intentionally untreated syphilis patients were allowed to suffer permanent and in some cases fatal sequelae.
5. The Lyme bacterium spreads to areas of the body that render this organism resistant to being killed by the immune system and by antibiotics, such as in the eye, deep within tendons, and within cells. The Lyme bacterium also has a very complex life cycle that renders it resistant to simple treatment strategies. Therefore, to be effective, antibiotics must be given in generous doses over several months, until signs of active infection have cleared. Because relapses have appeared long after seemingly adequate therapy, long term follow up, measured in years or decades, is required before any treatment regimen is deemed adequate or curative.
6. When administered by skilled clinicians, the safety of long term antibiotic therapy has been firmly established.

The very existence of hundreds of Lyme support groups in this country, and the tens of thousands of dissatisfied, mistreated and ill patients whom these groups represent, underscores the many problems that exist out in the real world of Lyme disease. I ask and plead with you to hear their voices, listen to their stories, and work in an honest and unbiased way to help and protect the many Americans whose health is at risk from what now has become a political disease. Thank you.

References

1-Website- Http://www.lyme.org

2-Website- Http://canlyme.com

3- Lab 257: the Disturbing Story of the Government' s Secret
Plum Island Germ Laboratory; by Michael Christopher Carroll;
Copyright 2004 by Michael Christopher Carroll; HarperCollins
[HC] pg 13;

4-All quotations in the article are from *Bad Blood: The Tuskegee
Syphilis Experiment,* James H. Jones, expanded edition (New York:
Free Press, 1993). Article taken from Information please database.
Pearson Education, Inc 2007

5-Written by John Drulle, M.D. in December, 1990 and reprinted
by the John Drulle, MD Memorial Lyme Fund, Inc. in 2006

6- Connecticut Attny. General' s Press release on May 1, 2008
http://www.ct.gov/AG/cwp/view.asp?a=2795&q=414284